AFFORDABILITY OF WASTEWATER SERVICE

Second Edition

Edited by John Mastracchio

2022

Water Environment Federation
601 Wythe Street
Alexandria, VA 22314-1994 USA
https://www.wef.org

ISBN: 978-1-57278-409-3

About WEF

The Water Environment Federation (WEF) is a not-for-profit technical and educational organization of 35,000 individual members and 75 affiliated Member Associations representing water quality professionals around the world. Since 1928, WEF and its members have protected public health and the environment. As a global water sector leader, our mission is to connect water professionals; enrich the expertise of water professionals; increase the awareness of the impact and value of water; and provide a platform for water sector innovation. To learn more, visit www.wef.org.

Seventh edition edited by John Mastracchio

With contributions by

Janet Clements
Joe Crea
Jon Davis
Zachary Green
Rachel Landman
Henrietta Locklear
Michael Matichich
Andrew Matsumoto
Eric Rothstein
William B. Zieburtz Jr.

Text originally prepared by the **Affordability of Wastewater Service Task Force** of the **Water Environment Federation**

David S. Hasson, PhD, *Chair*

Principal Authors:

Peiffer Brandt
Kees Corssmit, PhD
Eric Rothstein, CPA

Reviewers and Contributors:

Rachel Stevens
Alan Cherubin
Rich Cortes, PE
Bob Decker
Myron Olstein
Karen Pallansch
Steven M. Ravel, PE, BCEE

Under the Direction of the **Utility Management Subcommittee** of the **Technical Practice Committee**

Special Publications of the Water Environment Federation

The WEF Technical Practice Committee (formerly the Committee on Sewage and Industrial Wastes Practice of the Federation of Sewage and Industrial Wastes Associations) was created by the Federation Board of Control on October 11, 1941. The primary function of the Committee is to originate and produce, through appropriate subcommittees, special publications dealing with technical aspects of the broad interests of the Federation. These publications are intended to provide background information through a review of technical practices and detailed procedures that research and experience have shown to be functional and practical.

Contents

List of Figures

List of Tables

Preface

This publication is intended to provide an overview of the problem of affordability of wastewater utility bills and to offer specific considerations that elected officials, utility managers, and utility staff might take into account when addressing affordability issues in their communities. Some of these topics include defining and measuring affordability, highlighting what features might be included in an assistance program for those unable to pay for basic service levels, identifying the primary targets for an assistance program and determining eligibility, listing possible performance measures for providing this assistance, and offering some case studies from communities that have been leaders in attempting to tackle these difficult issues.

Affordability concerns affect customers, utilities, and the entire local community. Because of the relatively rapid increases in wastewater infrastructure costs and rates to address aging infrastructure both now and likely into the future, this topic is more timely and important than ever before.

The second edition of this Special Publication was edited under the direction of John Mastracchio.

Authors' and reviewers' efforts were supported by the following organizations:

Galardi Rothstein Group
Jacobs
One Water Econ
Raftelis

1

Introduction

Wastewater utilities are capital intensive, and major replacements, upgrades, or new regulations can put a serious financial burden on ratepayers—even if debt financing allows utilities to borrow money and pay for capital expenditures over an extended period. Over the past several decades, the cost of providing wastewater service has increased significantly because of challenges associated with aging infrastructure, climate change, stricter regulations, the desire for more intense levels of treatment, and population declines in urban centers. Government mandates for wastewater systems related to addressing combined sewer overflow (CSO) controls, nutrient loadings, and total maximum daily loads (TMDLs) have added to wastewater service costs. For example, according to the U.S. Conference of Mayors, operation and maintenance expenditures have increased approximately 4% annually from 1993 to 2017, an increase partially because of deferred capital expenditures. The needs associated with wastewater pipe repairs and replacements have been estimated at $18 per wastewater customer, and are projected to increase by an average of approximately 5% annually (American Society of Civil Engineers [ASCE], 2021). The American Society of Engineers estimated that the cumulative capital investment needed in United States water and wastewater infrastructure is approximately $3.27 trillion (ASCE, 2020), which will continue to place upward pressure on the cost of wastewater and water service.

Such spending has already begun to swell utility rates. According to the U.S. Bureau of Labor Statistics (U.S. BLS, 2021a, 2021b) and the U.S. Census Bureau (2021), increasing utility costs have resulted in the cost of wastewater services (combined with water and trash) to outpace the cost of inflation (Figure 1.1), raising concerns about their affordability, and the trend of rising utility rates is expected to continue. The effects on the personal finances of struggling low-income ratepayers may continue to be problematic as increases in utility spending—and, therefore, utility bills—outpace ratepayer incomes. This highlights a significant reason why affordability is important today.

1.0 DEFINITION OF AFFORDABILITY

Affordability is a concept that is easily understood but difficult to define precisely. Basically, it means that customers can pay their utility bills (and thereby meet utility needs) without undue hardship. It is an inherently subjective determination about what a utility customer can pay without excessive difficulty. However, utilities may need an objective definition that they can use when dealing with regulators, responding to ratepayers, making decisions on utility spending, or establishing customer assistance programs.

In this book, *affordability* is defined as the ability of households to pay for utility services without undue economic hardship or significantly downgrading their lifestyle below generally accepted minimum standards for health and welfare. Undue economic hardship refers to the need for fiscally

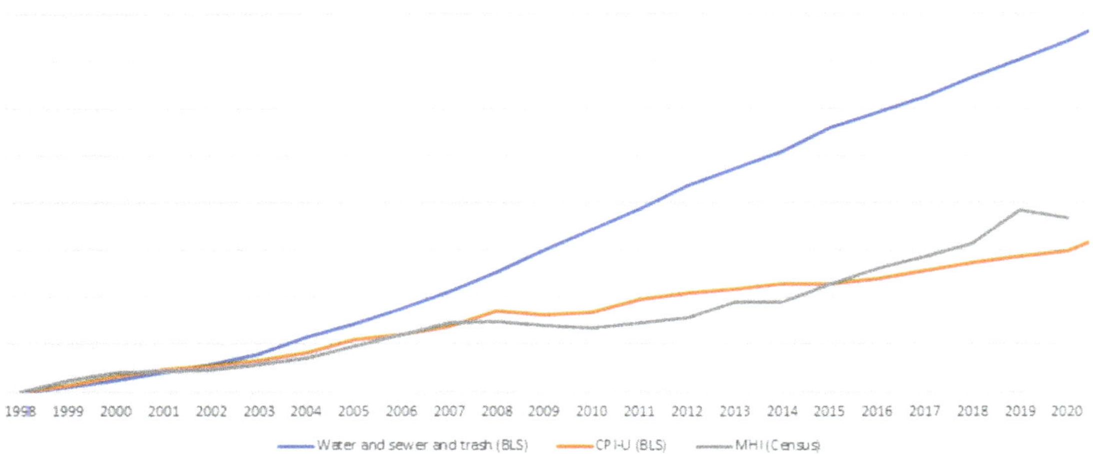

FIGURE 1.1 Utility Cost Increases (U.S. BLS, 2021a) Compared to Increases in Inflation (U.S. BLS, 2021b) and Median Household Income (U.S. Census Bureau, 2021)

challenged households to sacrifice other essential goods and services to pay their utility bills. Examples of economic hardship may include forgoing medically necessary prescriptions or doctor visits, sacrificing healthy meals, lacking the ability to fully pay for child care, essential transportation, or heating/energy services, or to cover rent or mortgage payments. Households that face water service shutoffs because of arrearages may be considered another example of undue economic hardship, as the loss of water services may in turn result in the loss of the habitability of their home or apartment (Raucher et al., 2019).

Although subjective, this definition clearly has broad applicability. For example, it would include a customer with a moderate income and extraordinary, ongoing medical expenses. For this customer, paying a utility bill might involve trade-offs with medical needs, food, or other essentials. An affordability definition based strictly on income would have excluded this customer. Depending on local circumstances, such as local cost of living differences, utility managers may have to "translate" this definition into more objective criteria to evaluate affordability or implement customer assistance programs.

2.0 PURPOSE AND OBJECTIVE

This book is designed to provide an overview of affordability from the wastewater services perspective, including measurement of affordability, regulatory considerations, approaches to addressing affordability (including customer assistance programs and eligibility considerations), funding of affordability programs, affordability program effects on the utility (including the identification of costs and benefits), program administration, and performance measures. Although framed from the wastewater services perspective, affordability issues apply to all water sector utilities, including wastewater, water, and stormwater, and the material contained in this book is equally applicable to these utilities. In addition, the book is intended to provide the reader with examples of programs that have been adopted and implemented by utilities to provide some real-world context to the topics discussed throughout these pages. The book also discusses the concept of equity and fairness in evaluating and addressing affordability.

Over the past decade, there has been a significant and growing body of knowledge regarding affordability that has appeared in the literature and applied in practice. This book is intended to provide a brief overview of this content rather than an in-depth treatise on the details of this body of knowledge. However, references to other sources of information that contain more detailed discussions of various affordability topics are referenced throughout this book.

3.0 REFERENCES

American Society of Civil Engineers. (2020). *The economic benefits of investing in water infrastructure: How a failure to act would affect the US economic recovery.*

American Society of Civil Engineers. (2021). *Infrastructure report card: Wastewater.* https://infrastructurereportcard.org/cat-item/wastewater

Raucher, R., Clements, J., Rothstein, E., Mastracchio, J., & Green, Z. (2019, April 17). *Developing a new framework for household affordability and financial capability assessment (FCA) in the water sector.* American Water Works Association, National Association of Clean Water Agencies, and Water Environment Federation. https://www.awwa.org/Portals/0/AWWA/ETS/Resources/DevelopingNewFrameworkForAffordability.pdf?ver=2020-02-03-090519-813

U.S. Bureau of Labor Statistics. (2021a). *Consumer price index for all urban consumers: Water and sewer and trash collection services in U.S. city average.* Federal Reserve Bank of St. Louis. Retrieved September 2021, from https://fred.stlouisfed.org/series/CUSR0000SEHG

U.S. Bureau of Labor Statistics. (2021b). *CPI for all urban consumers (CPI-U).* U.S. Department of Labor. Retrieved September 2021, from https://data.bls.gov/timeseries/CUSR0000SA0&output_view=pct_1mth

U.S. Census Bureau. (2021, September). Income and poverty in the United States: 2019, Table A-2. https://www2.census.gov/programs-surveys/demo/tables/p60/273/tableA2.xlsx

2

Understanding and Measuring Customer Affordability

This chapter will introduce a range of customer affordability metrics that are used to measure wastewater affordability in the water sector, the context in which they are used, and their limitations. Definitions of affordability vary because of the inherent subjectivity of the concept and the wide variety of circumstances involved (U.S. Congressional Budget Office, 2002). Some attempts have been made to establish community-based affordability standards, but experts recognize that what is affordable for a community or utility service area may not be affordable for some of its residents.

As noted in Chapter 1, affordability means that customers can pay their utility bills (and thereby meet utility needs) without undue hardship. Utility bills may be deemed unaffordable if they impair a customer's ability to pay for food, basic housing, normal medical needs, child care, and other essential items. So, when assessing the affordability of wastewater service, utility personnel must determine whether and how much of an "undue hardship" exists. Furthermore, affordability challenges are not just limited to one service, such as wastewater, but also apply to all utility services, including water, wastewater, and stormwater. Affordability is even more broadly defined as the ability to pay for *all* essential services and, as such, is applicable to more than the "one water" concept. However, in this chapter, understanding and measuring customer affordability is focused on the "one water" concept that encompasses all water-related services.

1.0 PURPOSE OF AFFORDABILITY METRICS

Affordability metrics may be calculated and reviewed for a range of different purposes, including as part of assessments of:

- Financial effects that might result from regulatory policies (on customers and/or utilities)
- Funding needs for grants or loans
- Wastewater service pricing
- Design of customer assistance programs (CAPs)
- Regional economic competitiveness comparisons
- Economic vulnerability or poverty studies

We can broadly divide these purposes into those that are policy related, which are typically driven by higher levels of government, and those that are rate related, which are more typically driven by local service providers.

1.1 Policy Context

The United States Environmental Protection Agency (U.S. EPA) has been prominent in the development of water and wastewater service affordability metrics. Perhaps the most well-known application of those metrics is found within U.S. EPA's Combined Sewer Overflow (CSO) Control Policy Financial Capability Assessment (FCA) guidance. In that case, affordability metrics were applied as one part of a set of financial capability metrics in an attempt to determine if CSO regulations would place an undue burden on a given utility over the period of the compliance schedule. It is important to note that the goal of this guidance is ultimately to measure an overall utility service area's ability to achieve regulatory compliance from a financial perspective, but not expressly to determine if the actual rates paid are affordable.

U.S. EPA also allows for application of affordability metrics in assessments of small drinking water system (SDWS) regulatory requirements, for which technology variances may be granted to communities with a population of fewer than 10,000 that are suffering from high affordability burdens.

One additional state-level policy context in which affordability metrics have been applied is as part of state revolving loan fund programs, which may award priority points to applicants found to suffer from affordability challenges. Under these programs, states administering U.S. EPA grants and loans for utilities develop their own affordability metrics, which may align, add to, or diverge entirely from U.S. EPA methodologies at each state's discretion.

1.2 Rate Context

Utility managers or finance directors may be interested in affordability metrics in a rate-setting context, as well to ensure that as budgets grow, rates remain affordable. Other utility applications might include measuring rate affordability to identifying households in need of assistance, or to inform the design of CAPs and measure their performance or effects over time.

Outside of the utility context, local policymakers may focus on essential service affordability, including wastewater services, for studies of regional economic competitiveness or economic vulnerability and poverty.

2.0 LIMITATIONS OF AFFORDABILITY METRICS

Each of the policy and rate contexts noted above suffers from data and other limitations that have received increased attention from utility experts, academics, and industry associations since the last edition of this book was published. Data and information challenges include, but are not limited to, the following:

- Consideration and available data on other essential service costs beyond the service under assessment
- Availability of spatial data sets that align with utility service areas
- Measurement of bills as a percentage of median household income (MHI) as opposed to lower, more vulnerable income levels

- Confusion over the distinction between metrics designed to assess utility financial capability and metrics focused on affordability for individual households
- Measurement at a point in time versus over time
- Consideration and available data on variable customer characteristics, such as household size or fixed incomes
- Variability in utility staff resources, analytical capabilities, and availability to participate in measurement

The balance of this chapter focuses on the history of affordability metrics and recently proposed innovations designed to overcome these noted limitations as best as possible.

3.0 U.S. ENVIRONMENTAL PROTECTION AGENCY AFFORDABILITY METRICS

The most commonly cited affordability guidelines or metrics are those issued by the U.S. EPA. This agency, which is responsible for regulating water and wastewater quality, has been evaluating affordability since at least the mid-1990s. The sections below detail the history of these guidelines, and how they have evolved.

3.1 History of Bills as a Percentage of Income and Thresholds

In 1995, the U.S. EPA prepared draft affordability criteria for infrastructure investments and other costs incurred to meet water quality standards. These criteria were intended as interim guidelines for use in evaluating the cost of implementing proposed water quality and pollution control systems, not for evaluating overall water system rates and charges. The guidance indicated that wastewater costs were affordable if the ratio of total annual charges to MHI was less than 1%. If the ratio was more than 2%, the costs placed an "unreasonable burden" on many households in the community and were therefore unaffordable (U.S. EPA, 1995).

If costs as a percentage of MHI was between 1% and 2%, the U.S. EPA recommended that utility staff evaluate several secondary criteria to determine the community's overall economic health and the proposed system's probable financial effect. These secondary criteria include:

- **Bond rating.** This indicator measures the community's creditworthiness and reflects its overall financial condition. However, not all communities have used debt financing, and of those that have, some have not had their bonds rated.

- **Overall net debt as a percentage of the full market value of taxable property.** This indicator measures the community's wealth by comparing its debt to the property values that may be required to support that debt.
- **Unemployment rate.** This indicator measures the community's current economic health, as well as the percentage of the population that may find rate increases unaffordable. A community's unemployment rate is considered above the national average if the difference between its rate and the national average is more than 1%.
- **Median household income.** This indicator measures a customer's general ability to pay for facility improvements. A community's MHI is considered to be above or below the state's MHI if there is more than a 10% difference between the two.
- **Property tax revenue as a percentage of the full market value of taxable property.** This indicator compares the community's funding capacity to its overall wealth.
- **Property tax collection rate.** This indicator measures how efficiently taxes are collected and, indirectly, how well the community's government functions.

Each criterion is rated "strong," "mid-range," or "weak," and then an average is calculated to determine the overall affordability rating.

The U.S. EPA also sets rate requirements for utilities that receive funding under its Construction Grants Program. It allows grantees to establish special low-income user charges at a fixed percentage of the normal rate for residents (Code of Federal Regulations, 2003). In this case, "low income" is defined as any residence in which the household income is less than the federal poverty level or defined as low income under state law or regulation. All other users must pay fees based on their proportional use of the system.

A 1997 U.S. EPA guidance manual on communities' ability to fund CSO control projects offered more guidance on determining the affordability of wastewater charges. It mirrored the agency's 1995 interim guidance for water quality standards (U.S. EPA, 1995) and framed it again as a two-step process. First, the guidance required a calculation of the cost per household as a percent of MHI, which is known as the residential indicator. Second, depending upon the results of the first calculation, the guidance required an evaluation of six additional criteria on a community's financial health, resulting in an assessment of the project's burden on the community (U.S. EPA, 1997).

The 1997 guidance manual also allowed for more information to be provided to support the financial capability assessments. Permittees were encouraged to include any supplemental information that might portray the community's financial capability more accurately. Using a matrix of results from the two-step evaluation, the community's financial capability could be categorized as "low burden," "medium burden," or "high burden." This categorization is then used to determine the implementation deadlines for CSO controls. The timing suggested by the guidance typically ranges from what U.S. EPA calls "normal engineering and construction practices" schedules up to 15 or possibly even 20 years. Keep in mind that this U.S. EPA book offers guidance; it does not mandate affordability determinations and the resultant suggested timelines for possible construction projects and programs.

As for a specific affordability threshold, this guidance manual indicated that investments in wastewater systems are typically affordable if the residential indicator equaled 2% of MHI or less (U.S. EPA, 1997). This community-based affordability standard does not automatically imply that all utility customers can afford the service, however; U.S. EPA acknowledges that some customers would still have difficulty paying their utility bills.

3.2 Updated U.S. Environmental Protection Agency Affordability and Financial Capability Assessment Guidance

In 1998, the U.S. EPA issued draft affordability criteria for small drinking water systems in which the affordability threshold was given as 2.5% of MHI (Federal Register, 1998). The agency also issued a 1998 report on affordability criteria for drinking water, which contained the following observations on affordability:

- There is a difference between willingness to pay and ability to pay. In areas where water was historically underpriced, a price increase might trigger an unwillingness to pay even though the price itself actually remains affordable. Price responsiveness should not be confused with inability to pay.
- Employment, income, and cost of living influence a person's ability to pay utility bills. Economies of scale of system operations may also affect affordability, depending on the circumstances.
- Alternate utility funding sources and rate structures can sometimes make utility bills more affordable.

- Affordability at the household level differs from affordability at the community level because of variations in income distribution. In practice, affordability is most often determined at the community level. (U.S. EPA, 1998)

An independent water system affordability study performed for the U.S. EPA states that an affordability threshold could range from 1.5% to 3.0% of MHI (International Consultants Inc., 1998). Within this range, a general target of 2.5% was selected based on several factors, including regulatory compliance. This criterion was intended to apply to one utility service (e.g., water), not multiple utilities (e.g., water and wastewater). In fact, most affordability thresholds are typically established for one service rather than the combined charges of multiple services.

The U.S. Small Business Administration convened the Small Business Advocacy Review Panel in 1999 to consider water quality regulations related to arsenic. The panel also addressed U.S. EPA's affordability threshold, concluding that it seemed to be higher than those used by other agencies, organizations, and various states (U.S. Small Business Administration, 1999). Panel members also suggested a less restrictive standard to allow for consideration of various circumstances so that individual states could make affordability determinations based on statewide criteria. The panel expressed concerns about a national affordability standard that did not consider multiple treatment processes, the variability of treatment costs, the variability of incomes, and other factors.

In 2002, the U.S. Congress directed the U.S. EPA to review its affordability criteria (Federal Register, 2002). The Affordability Review Panel of EPA's Science Advisory Board issued a report in December 2002 concluding that the affordability threshold concept had merit, but the threshold level should be lowered. It had never been exceeded even though many communities had difficulty funding the necessary utility infrastructure and could not afford the resulting utility bills (Science Advisory Board, 2002). The panel also requested formal guidelines for local variances and asked the agency to consider determining affordability on a regional or local basis instead of a national one. However, national guidelines helped promote fairness, the report noted, and U.S. EPA's approach had merit. The report also suggested that measures other than MHI might better reflect the effects on low-income households, but no alternatives were provided.

Another study seems to support the hypothesis that the affordability threshold guidance may be too high; the author found that the average household in the United States spends 1.5% of its income on both water and wastewater combined, which is well below the U.S. EPA guidance levels for affordability (Rubin, 2004). Furthermore, even in the state where the percentage of spending on water and wastewater services is the highest (West Virginia), an average of just 2.2% of income was spent on these services.

More recent studies and reports issued by the American Water Works Association (AWWA), the Water Environment Federation (WEF), the National Association of Clean Water Agencies (NACWA) (Raucher et al., 2019), the National Academy of Public Administration (NAPA) (Czerwinski et al., 2017), and the U.S. EPA's own Environmental Finance Advisory Board (EFAB, 2014) further built the case that the residential indicator was flawed for a number of reasons, including but not limited to the following:

- Median household income is a metric that does not account for the growing income inequality found in the United States today, so measuring affordability at that median level does not adequately assess the effects of rates for the most vulnerable households.
- The cost per household is narrowly defined by U.S. EPA but is not necessarily equivalent to the actual rates and consumption-based bill amounts experienced by customers, particularly for households of different sizes.
- Given that water sector rates are often linked (wastewater revenue requirements often include stormwater costs, and wastewater is often billed based on water consumption), it makes more sense for affordability to be measured for water sector service rather than for wastewater alone.
- Costs change over time and so affordability measurement should be integrated into a long-term cash flow model to show effects over time rather than only at one point in time at the beginning of a compliance schedule.
- Essential costs of living, such as food, housing, child care, health care, and utility costs of all kinds, should be considered when evaluating rate affordability for any one essential public service.

It is important to note that the U.S. EPA did come to accommodate some best practices advocated by the industry associations and others, such as permitting the submission of some supplemental information around costs of living and allowing for measurement that would account for geographic disparities in MHI (Ramseur, 2017). Still, in terms of how the measure was typically applied, and because of the limited capacity of some utilities to compile an extensive dossier on the level of burden that their community experienced, the guidance was generally thought to be overly narrow and poorly designed.

In February 2022, however, after concluding an extensive review process that considered submissions from water sector advocates, the U.S. EPA has taken some reformative action to address concerns. Changes to the latest CSO guidance's approach to affordability measurement are detailed below, though it is important to note that these are yet to be formally adopted as of this writing. The new draft guidance (U.S. EPA, 2022) provides the permittee an opportunity to add two additional means to communicate their level of financial capability, one focused on low-income burdens and the other on impacts to cash flows over time.

3.2.1 Lowest Quintile Income

The residential indicator is still included in the 2022 proposed guidance but it will be supplemented with the lowest quintile income (LQI) metric and a set of poverty-related metrics. These measures for the local service area will be compared to the same measures at the national level. The LQI metric is used because it typically represents demarcation between lower and middle class and the point at which households are not eligible for other forms of public assistance based on income. This leaves these "gap" households particularly vulnerable to affordability challenges for essential costs of living. Where the scoring identifies high levels of burden for the lowest quintile, permittees will then be required to document whether the community has considered all feasible steps to address impacts to the lowest quintile, such as the use of variable rate structures, CAPs, and applications for grants or subsidies from the Clean Water State Revolving Loan Fund program.

3.2.2 Total Water Sector Service Over Time

While the updated metrics still focus on wastewater costs, parallel residential indicator analytics and other supplemental information for drinking water costs are encouraged by the U.S. EPA in the updated guidance. Further, costs per household that include consideration of customer assistance programs, asset management costs, and stormwater management costs that were not specifically identified in the previous FCA for inclusion in the costs per household calculation are now specified. Finally, a cash flow forecasting model approach has been added, referred to in the guidance as an alternative that can be included to reflect the projected financial viability of the utility and the projected utility rates required to achieve the regulatory requirements over time. This alternative cash flow projection is intended to complement the income, poverty, and financial capability measures previously identified.

3.2.3 Poverty Measures

The new draft guidance (U.S. EPA, 2022) adds a poverty indicator that categorizes six measures of low-income status across three levels of burden (strong, mid-range, weak). These measures focus attention not only on residential indicator metrics but on more general economic vulnerability metrics that are widely available in federal databases. It also provides an indication of the prevalence of affordability challenges within a community by measuring the percent of households with incomes below established poverty thresholds.

The new draft guidance proposes to combine poverty and 20th-percentile income metrics and require a comparison with national statistics using a combined scoring or matrix approach. However, at the time of this writing, it was unclear which options identified in the draft guidance will be included in the final FCA guidance.

3.2.4 Household Types

Assessments of bills based on billed rates and consumption levels as a percentage of MHI and by household size were also included in the new draft guidance. These changes recognize that the old cost per household metric from the 1997 guidance did not reflect the actual rates experienced by customers, and that households of varying sizes may endure varying affordability burdens at the same income level.

4.0 ALTERNATIVE AFFORDABILITY METRICS AND THRESHOLDS

In recent years, water and wastewater practitioners have developed alternative affordability metrics (and associated thresholds) intended to help utility managers better capture the effect of water and wastewater costs on low-income households and to understand the nature of affordability challenges within their communities. These metrics include household burden indicators that compare the cost of water sector services to established income levels (and in some cases, account for the cost of other essential household expenditures), as well as prevalence indicators that measure the percentage of households within a community that may be facing affordability challenges. Table 2.1 provides a summary of alternative affordability metrics described in this section, as well as those defined by the U.S. EPA. Subsequent text describes each metric in additional detail.

4.1 Household Burden and Poverty Prevalence Indicators

A report prepared for the AWWA, WEF, and NACWA (Raucher et al., 2019) outlines a framework for assessing household affordability and financial capability within a federal regulatory context. This report proposed two

TABLE 2.1 Summary of Affordability Metrics

Metric Name	Equation	Affordability Threshold
U.S. EPA-based metrics (RI)[a]	Wastewater/water cost per household / MHI of service area	4.5% (2% wastewater, 2.5% water)
U.S. EPA LQRI[b]	Wastewater/water cost for lowest quintile household / LQI of service area	2% (high effect)
Household Burden Indicator (HBI)[c]	Basic water service costs (189 lpcd) / LQI[c] of service area	7% – 10% (high burden) >10% (very high burden)
Poverty Prevalence Indicator[d]	% of households earning < 200% FPL	>35% (very high burden) >20% (high burden)
WARi™	Average annual household bill (by census tract)[e] / MHI or LQI (by census tract)	2% (wastewater, MHI)
Teodoro Affordability Ratio for Lowest Quintile Income (AR_{20})[f]	Basic water service costs (189 lpcd) / Discretionary income of LQI	10% for 4-person hh[g] at 20th percentile income
Pacific Institute AR	Average water and wastewater bill for 163 lpcd / Discretionary income[h]	10% + for 4-person hh less than satisfactory; >30% unacceptable
Hours of Minimum Wage Equivalent[f]	Basic water service costs (189 lpcd) / Hourly minimum wage rate	8 hours for 4-person hh
Percentage of Household Income Spent on Housing	Selected monthly owner costs (or gross rent) / Household income	>30% (moderate burden) >50% (severe burden)

[a]RI = Residential indicator (U.S. Conference of Mayors et al., 2013; U.S. EPA, 1997)
[b]LQRI = Lowest quintile residential indicator (U.S. EPA, 2021)
[c]HBI is calculated for average household size; water service costs include costs for water, sewer, stormwater, assuming basic level of water use of 189 liters per capita per day (lpcd) (50 gpcd). The numeric values shown for the HBI threshold levels are preliminary, and subject to revision based on future empirical investigation (Raucher et al., 2019).
[d]FPL = Federal poverty level (Raucher et al., 2019)
[e]Calculates the weighted average financial burdens across all income levels, in all census tracts in a given utility's service area.
[f]Typically calculated for four-person household; includes water and wastewater costs for basic level of use of 189 lpcd (50 gpcd); discretionary income is defined as after-tax household income minus essential costs including shelter, health care, food, and home energy (Teodoro, 2018).
[g]hh = household
[h]Where discretionary income is household gross income minus expenses on shelter, health care, food, **transportation, telephone, laundry and cleaning**, home energy, and taxes. Items in bold were not included.

key metrics for assessing household affordability: the household burden indicator (HBI) and the poverty prevalence indicator (PPI).

The HBI reflects the economic burden of water services for low-income households. Specifically, the HBI considers annual water service costs (including water, wastewater, and stormwater) associated with a basic level of use (189 liters per capita per day [lpcd] or 50 gallons per capita per day [gpcd]) as defined by Teodoro (2018) for the average household size as a percent of the upper limit of the LQI. The PPI reflects the percent of the population living below 200% of the federal poverty level (FPL). The FPL provides a measure of income that corresponds to a minimum standard of living for different household sizes. This measurement is often criticized for being inadequate because the base data is outdated and does not account for geographic variations in the local cost of living. However, past studies suggest that, as a general rule of thumb, households must earn at least 200% of FPL to meet basic needs.

As detailed above, the U.S. EPA's updated FCA guidance incorporates similar metrics, including the LQI and the poverty indicator. However, the HBI differs from U.S. EPA proposed measures in that it explicitly includes costs for water, wastewater, and stormwater. It also bases costs on a level of basic use (189 lpcd or 50 gpcd), rather than average costs, which is arguably more relevant within the context of affordability and low-income households. To calculate the HBI, the authors of the AWWA, WEF, and NACWA report (Raucher et al., 2019) recommend using local utility rates to better reflect the reality of what customers need to pay to cover their basic needs.

Raucher et al., 2019 provides some guidance for establishing appropriate thresholds for the HBI and PPI, although the authors note that more empirical research is needed on this topic (e.g., HBI thresholds were developed by extrapolating the U.S. EPA's 4.5% of MHI to the upper level of the lowest income level). In short, water services are deemed to have a higher economic burden if total water costs for the average household size constitute a relatively large percentage of the LQI (i.e., greater than 7%) and a relatively large proportion of the community is composed of low-income households.

4.2 Weighted Average Residential Index

The Weighted Average Residential Index (WARi) was developed as an enhancement to EPA's residential indicator to account for differences in income levels, water use, and associated household bills across a service area. Specifically, WARi calculates the actual average utility bill within a census tract and compares the bill to household income levels (i.e., MHI,

LQI) within that tract. It then weights each tract based on the percentage of total households within the utility service area it comprises before summing up these percentages to calculate an overall weighted residential indicator or LQRI. The Weighted Average Residential Index offers a more accurate and refined assessment of household cost burden relative to the U.S. EPA's metric; however, it can require extensive analysis of utility billing data.

4.3 Household Water Service Costs as a Percentage of Discretionary Income

A limitation of many household burden metrics (e.g., HBI, WARI, and LQRI) is that they do not account for nondiscretionary expenses that low-income households face. Households struggling to meet basic needs face significant trade-offs in the allocation of their budgets. For example, small increases in rent or water and wastewater bills can adversely affect a low-income household's ability to pay for shelter, food, heat, and medical care (Raucher et al., 2019; United Way of Northern New Jersey, 2021). To better understand the effect of household water and wastewater bills within this context, researchers have developed metrics that compare household water service costs to discretionary income levels.

Teodoro (2018) developed the affordability ratio (AR_{20}), which compares the cost of basic water and wastewater services (at 189 lpcd or 50 gpcd; does not include stormwater costs) for a given household size to household discretionary income at the 20th percentile income level. In this case, discretionary income is defined as after-tax income minus costs for essential items, including shelter, health care, food, and home energy. The ratio is typically applied to a four-person household at the upper limit of the LQI but can be calculated for various household sizes and income levels. An AR_{20} threshold of 10% is contemplated in the research that introduced the metric, based on a preliminary distributional review of available data. The methodology for estimating the AR_{20} relies on extensive modeling of household expenditure data. However, in many cases, it can be approximated using publicly available data from the U.S. Bureau of Labor Statistics.

The Pacific Institute (Feinstein, 2018) proposes a similar metric, using 163 lpcd (43 gpcd) as the definition of basic water use. The institute defines discretionary income as gross household income minus expenses on shelter, health care, food, transportation, telephone, laundry/cleaning, home energy, and taxes. The Pacific Institute recommends thresholds for this indicator across satisfactory, moderate, marginal, and unacceptable affordability service levels. Specifically, with a "satisfactory" level of affordability, water and wastewater costs should account for no more than 10% of household discretionary income, while 30% would yield an unacceptable rating. The Pacific Institute recognizes that the affordability cutoffs are somewhat arbitrary but notes that it seems reasonable to infer that a household spending more than 30% of discretionary income would face unreasonable trade-offs with other necessary expenditures, especially when unanticipated large expenses arise.

4.4 Hours of Minimum Wage Equivalent

Teodoro (2018) also developed the hours at minimum wage metric, which measures the number of hours an individual earning the local minimum wage would have to work (before taxes) to cover basic household water and wastewater costs for one month. Like the AR_{20}, costs for this metric are based on a four-person household using 189 lpcd (50 gpcd). Research on the metric proposed a threshold of eight hours at minimum wage per month based on distributional analyses, suggesting that more than this indicates an affordability burden.

4.5 Percentage of Household Income Spent on Housing

The U.S. Census American Community Survey (ACS) contains data on the percentage of income that households spend on housing-related costs, including water and wastewater bills. The American Community Survey can be accessed online at https://www.census.gov/programs-surveys/acs. This data is available for both renter- and owner-occupied households. For renters, ACS calculates gross rent based on the total contract rent amount plus the estimated average monthly cost of utilities and fuels (if paid by the renter). For owners, ACS defines selected monthly owner costs as the sum of payments for mortgages and debts on the property, real estate taxes, insurance, and utilities and fuels. Households spending more than 30% of their income on housing are considered to have a moderate housing cost burden, while households spending 50% are considered to have a severe housing cost burden. While this metric excludes other basic expenses (e.g., food, taxes) and does not separately examine the effect of water and wastewater costs, the 30% and 50% thresholds are widely accepted as standard thresholds in government low-income housing programs and related research.

4.6 Additional Prevalence Measures

In addition to the PPI and poverty metrics included in the 2021 U.S. EPA FCA draft guidance, there are several additional data sources that can help utility managers better understand the prevalence of affordability challenges within their community. Examples of such data available from the U.S. Census American Community Survey (available at various geographic scales) include the percentage of households receiving public assistance and/

or Supplemental Nutrition Assistance Program (SNAP) benefits, and the percentage of households paying more than 30% of their income for housing (by income bracket).

United Way's ALICE Project provides additional insights on the percentage of households that struggle to meet basic needs. For every county in 21 states, ALICE (which stands for Asset Limited, Income Constrained, and Employed) has developed a Household Survival Budget, a basic budget that includes the cost of five essential items (housing, child care, food, transportation, and health care), adjusted for differences in cost of living and household types (e.g., number of people, children, and working adults). For participating states, ALICE reports the percentage of households earning less than the Household Survival Budget in every county (United Way of Northern New Jersey, 2021).

While there are no established thresholds for these broader socioeconomic/prevalence indicators, they can be compared to relevant state and national data; they can also be used to assess the level of need for a customer assistance program and, in the case of ALICE, help to set appropriate income thresholds.

5.0 UNDERSTANDING VULNERABLE POPULATIONS

As outlined earlier in this chapter, household wastewater affordability metrics were initially developed and applied within a regulatory context—specifically, to determine the ability of a community to comply with water quality–related mandates. However, as affordability concerns have continued to grow across the water sector, updated affordability metrics have allowed utility managers to better understand the effect of water sector costs and the affordability challenges that low-income households face. While these metrics are applicable within the regulatory compliance framework, they can also be used to help utility managers assess the need for CAPs, evaluate existing CAPs, and/or consider rate structure changes.

Within this context, it is important to go beyond the metrics outlined in this chapter to better understand the characteristics of vulnerable populations within a utility service area. For example, data from the ACS can provide insights on the number of renters and multifamily households in a utility service area who may pay for wastewater through their rental payment. Analysis of utility billing data can also provide insights into the typical household bills for households in low-income areas of a utility service area. As outlined in Chapter 3, this data should be used to inform utility decisions related to rate structures and CAP design, and can be instrumental in ensuring that utilities effectively target and communicate CAPs to reach those most in need of assistance.

6.0 CONCLUSIONS

Inherent in any definition of water affordability—at the local or national levels—are metrics for benchmarking and measuring it. Until relatively recently, the most common way of measuring affordability in the wastewater sector was to compare household wastewater costs to local MHI. The residential indicator was developed for the purpose of determining a wastewater system's ability to comply with federal water quality regulations. However, despite its limitations, it has served an informal benchmark for assessing affordability and continues to be widely applied in different contexts. On the drinking water side, national policy related to regulatory variances for small systems resulted in a similar metric and associated threshold of 2.5%.

In recent years, water and wastewater practitioners, as well as the U.S. EPA, have developed new metrics to help utility managers better understand and characterize affordability challenges. The new metrics still measure utility costs as a percentage of income, but focus on more vulnerable populations by using lower incomes and incorporating poverty metrics. In addition, they work to include total water sector service costs, measure effects over time, and reflect actual customer bills when looking at costs. These metrics are also designed to be incorporated into cash flow models that can serve as useful tools for utilities when setting rates, communicating with regulators, or designing customer assistance programs.

It must be noted that some have suggested that changes to the often-applied residential indicator metric do not go far enough, not least because even moving from MHI to the lowest quintile of income leaves room for households below that lower level that still cannot afford services. That is a valid criticism, particularly if one views water as a right that should not be denied to any human regardless of their financial means.

However, in the current environment in the United States, where consumption-based rate revenues fund the majority of utility operations and where the cost of essential needs do not scale lower with income, moving to ever-lower income levels for measurement would inevitably lead to an indication that almost no utility service is affordable for everyone, unless paired with rate design reforms or customer assistance programs. Indeed, those supporting more sensitive measurement often support such redistributive reforms, and there is a separate policy debate to be had there. Certainly, measurement of affordability at lower income ranges is sensible when seeking to design, target, and implement a CAP that is focused on the most vulnerable households only (see Chapter 3 for further discussion of CAPs).

Perhaps the biggest remaining challenge for affordability measurement is that there is no well-regarded federal benchmark of local costs of living that that can be universally applied to compare the cost of essential services to income and better inform affordability at the local level. Further, economically vulnerable households in the United States are concentrated in urban areas with pronounced income inequality and are disproportionately communities of color. As the United States moves beyond the recent COVID-19 pandemic, with a fresh perspective on economic vulnerability during the toughest of times, it is critical for the water sector to continue to focus on this challenging issue.

7.0 REFERENCES

Code of Federal Regulations. (2003). Grants for construction of treatment works. CFR, Part 35.2140(i), Title 40.

Czerwinski, S. J., Fretwell, E., Fosler, R. S., Lindsey, G., & Pagano, M. A. (2017, October). *Developing a new framework for community affordability of clean water services*. U.S. Environmental Protection Agency, National Academy of Public Administration. https://s3.us-west-2.amazonaws.com/napa-2021/studies/developing-a-new-framework-for-community-affordability-of-clean-water-servi/NAPA_EPA_FINAL_REPORT_110117.pdf

Environmental Financial Advisory Board. (2014, September). *EFAB analysis and recommendations on draft financial capability assessment framework*. U.S. Environmental Protection Agency. https://www.epa.gov/sites/default/files/2015-10/documents/municipal_fca_framework.pdf

Federal Register. (1998, August 6). Announcement of small system compliance technology lists for existing national primary drinking water regulations and findings concerning variance technologies, vol. 63, no. 151, section A3; National Level Affordability Criteria, pp. 42, 46.

Federal Register. (2002, August 20). Meeting of the Small Systems Affordability Working Group of the National Drinking Water Advisory Council, vol. 67, no. 161, Supplemental Information, p. 62,718.

Feinstein, L. (2018, September). *Measuring progress toward universal access to water and sanitation in California: Defining goals, indicators, and performance measures*. Pacific Institute. https://pacinst.org/wp-content/uploads/2018/09/Measuring-Progress_Pacific-Institute_Sep-2018.pdf

International Consultants, Inc. (1998). *National affordability criteria under the 1996 amendments to the Safe Drinking Water Act*. U.S. Environmental Protection Agency.

Ramseur, J. L. (2017, March). *EPA policies concerning integrated planning and affordability of water infrastructure*. Congressional Record Service. https://www.everycrsreport.com/reports/R44223.html#Content

Raucher, R., Clements, J., Rothstein, E., Mastracchio, J., & Green, Z. (2019, April 17). *Developing a new framework for household affordability and financial capability assessment (FCA) in the water sector*. American Water Works Association, National Association of Clean Water Agencies, and Water Environment Federation. https://www.awwa.org/Portals/0/AWWA/ETS/Resources/DevelopingNewFrameworkForAffordability.pdf?ver=2020-02-03-090519-813

Rubin, S. J. (2004). *The cost of water and wastewater service in the United States*. National Rural Water Association.

Science Advisory Board. (2002, December 30). *Affordability criteria for small drinking water systems: An EPA Science Advisory Board report*. U.S. Environmental Protection Agency.

Teodoro, M. P. (2018). Measuring household affordability for water and sewer utilities. *Journal of the American Water Works Association*, *110*(1), 13–24.

United Way of Northern New Jersey. (2021). Meet Alice. https://www.unitedforalice.org/

U.S. Conference of Mayors, American Water Works Association, & Water Environment Federation. (2013). *Affordability assessment tool for federal water mandates*. Stratus Consulting. https://www.awwa.org/Portals/0/AWWA/ETS/Resources/AffordabilityAssessmentTool.pdf

U.S. Congressional Budget Office. (2002, November). *Future investment in drinking water and wastewater infrastructure*, Appendix C.

U.S. Environmental Protection Agency. (1995). *Interim economic guidance for water quality standards workbook*, section 2.3.

U.S. Environmental Protection Agency. (1997). *Combined sewer overflows—Guidance for financial capability assessment and schedule development*. Office of Water, Office of Wastewater Management.

U.S. Environmental Protection Agency. (1998). *Information for states on developing affordability criteria for drinking water*. Office of Water.

U.S. Environmental Protection Agency. (2021, January). *2021 financial capability assessment guidance*. https://www.epa.gov/sites/default/files/2021-01/documents/2021_fca_guidance_-_january_13_2021_final_prepub.pdf

U.S. Environmental Protection Agency. (2022, February). *Proposed 2022 clean water act financial capability assessment guidance*. https://downloads.regulations.gov/EPA-HQ-OW-2020-0426-0071/attachment_1.pdf

U.S. Small Business Administration. (1999, June). *Report of the small business advocacy review panel on EPA's planned proposal of the national primary drinking water regulation for arsenic*, section 9.2.2. Office of Advocacy. https://www.sba.gov/sites/default/files/files/Report_Drinking_Water_Regulation_Aresnic_06-04-1999.pdf

3

Approaches to Addressing Affordability

This chapter identifies the range of approaches that utilities; other local, state, and federal government agencies; nongovernmental organizations (NGOs); individual customers; and other entities can use to address the significant and continuing affordability challenges identified in this book. During the past dozen years, many utilities have adopted new customer assistance programs (CAPs) for low-income customers or have expanded/enhanced existing programs based on components such as bill discount programs and emergency assistance programs that have been in existence in the water sector for many years. In addition, new options have been added to the array of previous options, such as aligning charges for service directly with income for low-income residents, as was done in the Philadelphia example profiled in Chapter 8. Some of the approaches identified in this chapter, such as federal government subsidies patterned after the Low Income Home Energy Assistance Program (LIHEAP) and embedding human health or comparable affordability-related considerations directly into the cost allocation and rate-setting process, represent new ideas and approaches that are untested in actual practice but have gained traction and the interest of some finance practitioners in the water sector because of the perceived need for more robust or more efficient methods to complement the options currently in practice.

1.0 POLICY CONSIDERATION: WATER AS A HUMAN RIGHT

The focus on expanding existing approaches to affordability and developing new ones has emerged in response to the increased cost of meeting regulatory requirements and the asset-management needs of renewing and rehabilitating aging system assets. In addition, the focus on affordability in recent years has also resulted from consideration of environmental justice and social equity by some water and wastewater service providers. Throughout

North America and, indeed, globally, there is a heightened awareness that access to certain basic levels of water and wastewater service is essential to supporting human life at optimal levels of health and sanitation. There is also an increased awareness that while water and wastewater utilities can play an important part in addressing the affordability challenges necessary to provide essential water and wastewater service to all, ultimately addressing the affordability challenge will require shared recognition of responsibility and action by a variety of actors.

When the American Water Works Association (AWWA) and Water Environment Federation (WEF) sponsored a two-day symposium on affordability in 2018, Figure 3.1 was developed to frame some of the primary actions that government agencies, utilities, and social service agencies could take to help address the affordability challenge. These included direct-bill assistance, rate design, shutoff forbearance, education, and enhanced social services support, among other options.

2.0 IMPORTANCE OF MANAGING TOTAL UTILITY COSTS AS AN ELEMENT OF PROVIDING AFFORDABLE SERVICES

Sound water and wastewater utility management includes managing the total life-cycle cost of service so that it is cost efficient for customers. The water sector has recognized that one key strategy utilities can use to help make service affordable for all customers is to implement best practices that increase operating efficiencies, promote conservation, allocate resources properly, take advantage of regionalization/economies of scale, and manage capital and financing costs in order to reduce the total annual revenue

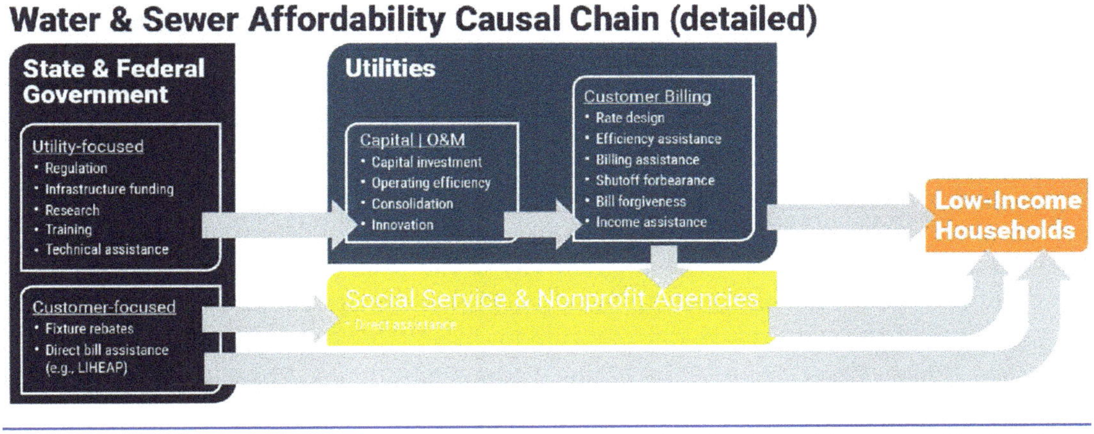

FIGURE 3.1 AWWA–WEF Illustration of Interconnected Roles for Addressing Affordability

requirements that need to be borne by all customers (AWWA, 2018). Some examples of how state and federal governments can collaborate with utilities to help reduce infrastructure financing costs include the long-standing existence of federally seeded state revolving fund loans, which provide low-interest loans to many utilities, and the relatively recent emergence of low-interest loans through the Water Infrastructure Finance and Innovation Act loan program administered by the U.S. Environmental Protection Agency (U.S. EPA). Recent federal allocations of funding for water and wastewater infrastructure, along with crisis assistance to customers through the new Low-Income Household Water Assistance Program (LIHWAP), are examples of how the federal government may assist with affordability nationwide by lowering the cost of wastewater service provision. Over the long term, federal and state action may be needed to help eventually solve issues with the high cost of wastewater service.

Given the considerable capital and operating needs of providing safe, quality water and wastewater services, many utilities that fully implement best management practices to the total life-cycle costs of providing service find that they have a significant number of low-income customers who cannot afford to pay for basic levels of water and wastewater service to support essential needs for human life. That is why WEF, AWWA, utilities, state and national government agencies, and other industry leaders support the implementation of additional measures, such as those identified in this chapter, to help make essential water and wastewater service available to all at affordable prices.

3.0 MECHANISMS FOR ADDRESSING AFFORDABILITY

This section provides a brief description of mechanisms currently being used by some wastewater and water utilities to address affordability for low-income customers, as well as some additional mechanisms currently in development and under discussion to address affordability. These mechanisms include CAPs and utility rates designed to more directly address affordability. Customer assistance programs are designed to provide assistance and relief to low-income and other disadvantaged customers, such as elderly and physically challenged customers who are unable to pay their utility bills because of affordability limitations.

Over the past several years, there have been a number of research studies and journal articles that provide more detailed descriptions of the mechanisms described here, including a compendium of CAPs released by the U.S. EPA in April 2016. Titled *Drinking Water and Wastewater Utility Customer Assistance Programs*, this compendium was the first survey of CAPs in all

50 U.S. states and was supported by WEF and a number of other industry associations.

The selection of specific strategies and mechanisms for addressing the affordability challenges within a specific utility system's customer base should be guided by a number of factors including, but not limited to:

- The severity of the affordability challenge, including the number of customers facing challenges in paying for essential levels of water and wastewater service and the severity of the need;
- Community values and support from other customers and community stakeholders to provide assistance to low-income customers;
- Availability of supporting resources, such as community service agencies, to help qualify and support the affordability programs; and
- Availability of administrative means to provide financial relief to customers needing assistance, including hard-to-reach customers.

Resources such as the Environmental Justice Screening and Mapping Tool (EJSCREEN) developed by the U.S. EPA (2020) and industry association guidance documents, such as the *Affordability Assessment Tool for Federal Water Mandates* (Stratus Consulting, 2013) developed by the U.S. Conference of Mayors, AWWA, and WEF, are available to utilities to support evaluation of these considerations in developing a strategy for addressing affordability.

3.1 Customer Assistance Programs

This section describes mechanisms that are primary elements of CAPs that are utilized by some utilities in the wastewater and water sector.

3.2 Bill Discounts

Bill discounts are often a primary component of a CAP. A bill discount is a tool to help ratepayers with their water and wastewater utility bills. Bill discounts are targeted to households with chronic affordability issues (i.e., where ongoing household income is insufficient to pay utility bills plus other necessities for living, such as food and shelter). This assistance can take a number of forms, all of which reduce a ratepayer's liability for water and wastewater services rendered. They enable utilities to effectively target beneficiaries while managing program costs. As detailed below, there are a number of types of bill discounts including:

- Fixed-dollar-amount discount
- Percentage-of-bill discount

- Percentage-of-bill discount up to a maximum
- Percentage-of-bill discount graduated by income level

While not in widespread use, another form of bill discount that has been tested in limited circumstances is providing discounts from outside the utility's revenue stream through federal or state bill discount subsidies.

3.2.1 Fixed-Dollar-Amount Discount

A fixed-dollar-amount discount provides qualifying ratepayers with utility bills that are reduced by a specific dollar amount, irrespective of the current balance, any past-due balances, or ratepayer income level. This method is simple, is relatively easy to administer, and has predictable costs, but it is largely insensitive to specific ratepayer needs. For example, it does not allow program staff to distinguish between a large, water-efficient household that has significant burdens and a small household that uses water less efficiently.

3.2.2 Percentage-of-Bill Discount

A percentage-of-bill discount provides qualifying ratepayers with utility bills that are reduced by a given percentage, irrespective of the balance due, ratepayer income, or resulting revenue implications. This method is more responsive to ratepayers' needs than the fixed-dollar-amount discount, but it is still insensitive to income level. This method also is relatively simple and easily administered, but its costs are less predictable, and it does not promote conservation.

3.2.3 Percentage-of-Bill Discount up to a Maximum

This method caps the percentage discount, thereby making program costs more predictable. It also provides further incentive for ratepayers to manage their water consumption.

3.2.4 Percentage-of-Bill Discount Graduated by Income Level

A percentage-of-bill discount that is graduated by income level provides qualifying ratepayers with utility bills that are discounted; the discount percentage varies based on the ratepayer's income level. Lower-income ratepayers are eligible for a larger discount than those with somewhat higher incomes. This method is more tailored to the needs of individual ratepayers, but it has substantially more administrative requirements and its costs are less predictable.

3.2.5 Federal- or State-Subsidized Bill Assistance

Most existing bill discount programs in the wastewater and water sector are funded by other utility system customers through their rates and charges, or through voluntary customer contributions such as bill round-up programs. In contrast, provision of necessary winter heating assistance to low-income customers is addressed through the federally funded LIHEAP. Recent federal legislation has authorized pilot programs for federal bill subsidies for low-income customers in the water and wastewater sector. Further, during the 2020 COVID-19 pandemic, the American Rescue Plan Act (ARPA) set aside $50 million in federal monies to provide assistance to low-income water and wastewater customers to help pay their bills; in addition, states, counties, and municipalities were allowed to use a portion of the $350 billion in assistance provided to local agencies through ARPA to provide additional assistance to water and wastewater customers.

3.3 Emergency Assistance

Some CAPs include an element for emergency assistance to provide temporary financial relief for customers who are unable to pay their wastewater and water bills because of temporary circumstances such as unemployment, health emergencies, or a change in family status. Crisis vouchers and bill forgiveness are examples of such program elements.

3.3.1 Crisis Vouchers

Sometimes ratepayers have one-time crises that limit or prohibit them from paying water and wastewater bills promptly; crisis vouchers enable eligible ratepayers to do so. To qualify, ratepayers must provide evidence of the crisis. They then are typically not eligible for another voucher until some predetermined time has lapsed (e.g., once per year). From an accounting perspective, crisis vouchers are considered an alternative means of payment. Crisis vouchers are used sparingly and often are combined with more permanent bill-assistance programs because the related administrative and implementation requirements to prevent abuse of the voucher system can be substantial.

3.3.2 Bill Forgiveness/Write-Offs

Write-offs of outstanding ratepayer liabilities provide relief by erasing some or all of an eligible ratepayer's outstanding account balances. They may be structured to eliminate interest charges on outstanding account balances, past-due fees, or the entire outstanding balance. This option may be provided for an irregular event occasioned by a ratepayer's unusual or particularly difficult situation. Few water and wastewater utilities include bill forgiveness as a component of a CAP, although this option is used more frequently by regulated energy utilities. Under this option, customers enrolled in a CAP are eligible for a write-off of a portion of their debt in exchange for making a timely monthly payment of bills. Bill write-offs typically have relatively substantial administrative requirements (e.g., determining eligibility and making appropriate adjustments to customer account records and associated accounts receivable).

3.4 Water Audits and Plumbing Repair Programs

Some utilities have found it useful and cost efficient to help low-income customers reduce their water usage, and therefore reduce the portion of water and wastewater bills that are based on variable usage of water. These program elements can take several forms, including:

- Education programs
- Conservation programs
- Leak-repair and plumbing-fixture replacement programs

3.4.1 Education Programs

Education programs help ratepayers manage their water consumption by modifying their water use practices. Such programs may be directed at low-income ratepayers or all residential ratepayers.

3.4.2 Water Audit and Conservation Programs

Conservation program personnel evaluate (audit) individual ratepayer practices and subsidize the installation of water-efficient devices (e.g., via rebates). Such programs may be directed at low-income ratepayers or all residential ratepayers.

3.4.3 Leak-Repair and Plumbing-Fixture Replacement Programs

Low-income ratepayers often have relatively high water-consumption patterns because of damaged and leaking plumbing and old, inefficient fixtures. Repair and replacement programs can solve this problem by helping eligible ratepayers fix leak-prone plumbing and install low-flow plumbing fixtures. This assistance may be offered to a limited number of ratepayers, since the related administrative costs (e.g., eligibility qualification, contractor management, and insurance) can be substantial. In some cases, utilities hire plumbers and offer to provide low-income customers assistance in repairing leaks identified through water audits up to a maximum equivalent dollar value; in other cases, utilities help to arrange low-interest loans to help low-income customers afford plumbing and related repairs that can help to reduce water usage and therefore the variable portion of the water or wastewater bill.

3.5 Payment Plans and Related Financial Counseling

Some utilities find that it is useful to provide counseling, payment plans, and related forms of assistance to help customers address cash flow and other financial issues that can help increase the ability of customers to pay their wastewater and water bills.

3.5.1 Financial Counseling and Referral

Program staff sometimes provide ratepayers with information and advice on managing their overall financial resources and regularly paying utility bills. This may include referrals to organizations such as social service agencies that can provide help with payment relief or debt restructuring, or can offer extensive financial counseling related to the cumulative effect of paying for the full range of necessary public services including winter heating and electric bills, food and shelter, and other necessities for living that households require.

3.5.2 Payment Plans

Payment plans can spread utility bill payments over several months to help ratepayers address substantial past-due balances. These plans may be structured with or without accrual of interest and administrative charges, may vary in duration, and may be offered under various eligibility provisions. These payment plans often include provisions in case the ratepayer defaults, because these plans are typically offered to ratepayers with substantial past-due balances. Such provisions may include placing property liens or shutting off services.

3.5.3 Disconnection Moratoria

Some utilities establish disconnection moratoria as an alternative form of ratepayer assistance. This tool assures ratepayers that they will have continuous service irrespective of account status; but to be effective, another method of collecting on past-due account balances is needed. Such methods may include placing liens on property or submitting accounts receivable to collection agencies. During the COVID-19 pandemic in 2020, most wastewater and water utilities in the United States placed some form of temporary moratoria on disconnecting customers from essential water and wastewater service as millions of families across the country faced severe financial hardship resulting from the loss of jobs with the shutdown of much of the economy. A portion of the national pandemic stimulus legislation was dedicated to enabling states to provide assistance to households whose pandemic-related financial hardship contributed to their challenges in paying their wastewater and water bills.

3.6 Addressing Affordability Directly in Cost Allocation and Rate Structure Programs

Addressing affordability issues directly in the cost of service and the rate structure development process, rather than using after-the-fact options such as bill discounts described above, is an option that some utilities have employed. This includes a number of rate structure–related approaches that have been in use for some time, such as:

- Lifeline rate options
- Maximum quantity subject to charges
- Inverted rate blocks

It also includes some emerging options such as:

- Area-based rates
- Income-based rates
- Addressing affordability directly in the cost-of-service/cost allocation process

3.6.1 Lifeline Rate Options

Lifeline rate options extend subsidized water and wastewater services to eligible low-income ratepayers. The subsidies could apply to all volumes of service, or they may be limited to specified volumes. To qualify, ratepayers must provide evidence of their special status, which may include low income, age, and/or physical challenges. This option imposes limited, predictable costs on the utility, but often requires significant implementation and outreach efforts. For purposes of this discussion, "lifeline rate options" are distinguished from a form of inverted block rate structure wherein initial volumes of use are priced at subsidized rates for all customers within the customer class subject to this rate option. This distinction is associated with the qualification requirements of lifeline rates.

3.6.2 Maximum Quantity Subject to Charges

Another affordability-oriented rate structure option involves capping the volumes subject to charges, so eligible low-income ratepayers' charges will not exceed a defined maximum within a given period. To qualify, ratepayers must provide evidence of their low-income status. This option imposes limited, predictable costs on the utility, but often requires significant implementation and outreach efforts.

3.6.3 Inverted Rate Blocks

Inverted rate blocks typically apply to all customers in an established customer class and involves establishing increasing unit charges for various ranges of water volumes. This option is relatively uncommon in the wastewater sector because ratepayers have relatively little control over their discharge volumes, and peak flows typically are associated with wet weather. However, inverted rate blocks can make wastewater service more affordable by establishing low unit prices (e.g., below cost of service) for the lowest range of volumes. It also can promote water use efficiency and, combined with conservation programs, may serve other, perhaps primary, community objectives. Converting to this rate structure requires careful analysis of a given customer class's consumption patterns to predict associated revenue effects, but once established with adequate billing-system reporting, this is a relatively easy form of assistance to implement and sustain.

3.6.4 Area-Based Rates

This is a modification of the traditional approach to customer classification, in which users are grouped based on service characteristics (e.g., wastewater flows and strengths). In this case, rates are based on geography (e.g., low-income zip codes) to make water and wastewater service more affordable in economically disadvantaged areas. This method has some disadvantages: it is imprecise, and does not necessarily uphold cost-of-service principles. In addition, it may not be allowed under specific state statutes. On the other hand, it may complement a community's economic development and environmental justice and social equity objectives. Area-based rates are also a relatively easy form of assistance to implement and sustain. In addition, cost-of-service analyses may be able to be used to cost-justify a differential in rates between inside- and outside-jurisdictional areas. If the inside-jurisdictional customers have lower incomes than outside-jurisdictional customers (e.g., an urban utility serving suburban areas), implementing rates that reflect such a cost differential could be beneficial for affordability.

3.6.5 Income-Based Rates

This is a modification of the traditional approach to rate-setting in which the specific bills for low-income customers are based on a household's income level. The Philadelphia case study included in Chapter 8 is the first example of implementation of this approach to rate-setting in the water industry. This method does not necessarily uphold cost-of-service principles but, like area-based rates, may complement a community's environmental justice and social equity objectives. Based on the City of Philadelphia's experience with this method for addressing affordability, there are considerable administrative costs associated with both implementing and sustaining this approach to addressing affordability.

3.7 Addressing Affordability Directly in the Cost-of-Service Cost Allocation Process

There is a growing dialogue among rate and financial researchers and practitioners in the water and wastewater sector regarding benefits to more directly addressing affordability in the cost-of-service cost allocation and rate-setting process, rather than addressing affordability after the fact based on cost-of-service analyses that do not factor in these considerations (see Giardina, 2019; Rothstein et al., 2021; Teodoro, 2018). These and other industry papers envision a methodology in which public health might be included as a separate factor to which costs are allocated based on the requirements to meet the basic human health needs of a community's residents and workers. While no communities have implemented such systems at the writing of this book, there are proactive efforts within the industry to develop guidance documents, example calculations, and other tools that would enable communities that find these methods appealing to implement such approaches to affordability.

3.8 Other Considerations

Rural and tribal communities face unique challenges when addressing customer affordability, and some affordability tools may be particularly applicable to assist them. Reducing the cost of wastewater services is one of several strategies to resolve affordability issues, and for rural utilities, collaborative service delivery, sharing resources, and even consolidation could be helpful. Rural and tribal communities may consider collaborating with surrounding utilities to lower operational costs. This in turn can lower customers' utility bills or provide additional funds for a customer assistance program. Collaboration and partnerships can range from specific contracts and shared services to completely consolidating finances and management.

There are many methods of collaboration including sharing equipment, treatment and delivery of water services, emergency planning, infrastructure and capital projects, and overall management and administration that can improve affordability of water for rural and tribal communities.

Additionally, there are multiple grants that may be available to rural and tribal communities with a focus on technical and financial assistance. One such grant is the Water and Waste Disposal Technical Assistance and Training Grant administered through the U.S. Department of Agriculture. Through this grant, rural and tribal communities can receive technical assistance and training from nonprofits that specialize in improving operation of water and waste facilities in rural areas. The U.S. Department of Agriculture has additional grants through the Rural Utilities Service Water and Environmental Program to provide technical assistance and financing for rural drinking water and waste disposal systems (U.S. Department of Agriculture, n.d.).

4.0 WHO IS ELIGIBLE FOR ASSISTANCE?

Utility management, regulators, and policymakers may operate on the assumption that affordability is only a concern for low-income customers, but any customer who has to make tough choices about which bills to pay because of a lack of financial resources is experiencing an affordability challenge. As discussed in this chapter, a common approach to providing customer assistance is to base assistance solely on income level. Income level–based qualification for assistance is relatively simple relative to other measures, but because other factors are relevant, alternative qualifications for assistance should be considered. These factors may include other essential or extraordinary expenditures, measures of wealth as opposed to income, temporary versus long-term financial limitations, and debt.

The definition of affordability itself may present a challenge when determining eligibility criteria for customer assistance. As discussed in Chapter 2, utilities and regulators define affordability differently, and refined definitions of affordability are emerging. When utilities discuss affordability, they typically are referencing the percentage of their customers who have difficulty paying wastewater bills. This reference to affordability is household or customer affordability. Regulators, on the other hand, typically discuss affordability in terms of an entire community's ability to pay for mandated improvements. From the perspective of designing CAPs, household affordability is the relevant measure.

Differing community circumstances, priorities, and social values may cause utilities to target different customers in their assistance programs.

Many CAPs target low-income customers generally, but some also have low-income senior citizen discounts or other programs.

Before adopting a particular approach to customer assistance, it is useful for utility policymakers to gather data and evaluate the issue of affordability in depth in their own communities to identify the criteria defining customers needing help, how best to reach these customers, and effective and legal ways to assist them. With respect to reaching customers, many utilities have found that based on analysis, low enrollment rates in their programs persist, compared with the numbers of customers who are likely eligible. Thus, both effective engagement and lowering barriers to enrollment should be considered together as a program is designed. Furthermore, utilities that adopt reciprocal eligibility with existing programs such as Temporary Assistance for Needy Families (TANF) and the Supplemental Nutrition Assistance Program (SNAP) may be able to streamline enrollment processes for their customers. As this chapter has described, utilities have taken a variety of approaches to foster affordability, but not all utilities are allowed the same options because state laws, charters, and other legal requirements vary and may restrict utilities from implementing CAPs. A good starting point for policymakers is the 2017 publication *Navigating Legal Pathways to Rate-Funded Customer Assistance Programs* (University of North Carolina Environmental Finance Center, 2017), which documented legal frameworks at that time. It should be noted that even if CAPs are allowed for utilities, providing assistance to certain groups wholesale may introduce the risk of legal challenges. Groups that have been targeted in the past are discussed below.

4.1 All Residential Customers

As described in the previous sections, approaches that are not administered programs per se, can be effective mechanisms for addressing affordability. For example, an inverted rate block structure seeks to provide a lower rate for essential water usage to all residential customers, such that these customers can afford the minimum level of service needed for essential purposes. Water utilities may do this by offering a minimum volume of water at a price below cost of service or at no volumetric charge. This approach is less common among wastewater utilities for various reasons, including lack of water-meter data, more use of flat-charge structures, and less use of tiered-rate structures compared to water utilities. An advantage of this approach is that it is easy to administer; such passive "assistance programs" involve almost no administrative time or cost—except the initial rate setup and periodic updating. A disadvantage of the approach is that it is not targeted: customers who may not need the benefit of lower-cost water for essential usage receive it. This may result in higher overall program costs, generating a cycle that exacerbates the affordability problem for those in need. Managing the total cost of wastewater service is also a key approach. While lowering the total costs of providing service may benefit all customers indiscriminately, it is an essential long-range approach to resolving affordability.

4.2 Low-Income Customers

Many assistance programs rely on an income criterion to determine customer eligibility: if a customer's income is at or below a specified level, then that customer is eligible for the program's benefits. The intent is to target customers who struggle to pay, using income level as the indicator of need because it is seen as the most common overriding factor affecting a customer's ability to pay the utility bill. Often, income is adjusted for household size, as is the case with programs that rely on federal poverty guidelines to set their eligibility thresholds. Some programs include other important factors, such as extraordinary medical expenses.

The advantage of an income-based program is that it focuses on customers who likely need assistance. A disadvantage of the approach, however, is the administrative time and costs involved in verifying customer eligibility. Utilities require staff or contractors to verify income levels, which is not a one-time review but an ongoing process because customers' incomes change over time. Eligibility determination may also involve verifying the number of residents in a household (and their incomes). On the other hand, some utilities have streamlined administration by accepting a determination from another program or programs as proof of meeting its income eligibility criterion. For example, a customer enrolled in another local or state assistance program that provides aid could qualify automatically for the utility's program.

Of note, income-based assistance programs tend not to distinguish between income and wealth. This has the effect of broadening program eligibility to customers who may be low income but because of their assets, are not making trade-offs in order to pay their wastewater bills. But, as with verifying income, verifying wealth (or the lack of it) represents an administrative cost, and therefore many programs avoid this added layer of complexity.

4.3 Senior Programs

Many sectors of the U.S. economy offer discounts to senior citizens. Such discounts are based on the premise that these customers are no longer working, have fixed (limited) incomes, and therefore cannot afford high or rising prices. Wastewater utilities are no exception.

An age-based criterion alone is easier to implement than an income criterion because age is easier to verify and does not require subsequent reverifications (i.e., eligible customers do not become younger). A criticism of basing an assistance program solely on age, however, is that senior citizens who do not need the benefits receive them anyway. To combat that issue, some programs require that customers meet both age and income thresholds in order to qualify for assistance. If an income criterion is added, eligible seniors become merely a subset of low-income customers. These customers differ from others only in terms of ongoing program administration design. These customers may be good candidates for less frequent income verification by the utility given that their incomes may be more stable or fixed than other customers' incomes.

4.4 Owners/Renters

In the wastewater sector, customer assistance programs most often focus on residential customers who are direct customers of the utility, while few provide assistance to those living in rental units or multifamily dwellings, or to businesses. Residential water and wastewater users who are not directly billed customers of utilities are sometimes termed "hard-to-reach customers." A 2017 study for the Water Research Foundation (WRF) titled *Customer Assistance Programs for Multi-Family Residential and Other Hard-to-Reach Customers* found that hard-to-reach customers made up over a fifth of all households and 40% of low-income households (Morin, 2017).

Residents of rental units and multifamily dwellings are often not utility customers; their landlords are. They may struggle with the high cost of housing, including utilities, but these bills may be a part of their rent. It may be a challenge for the utility to reach and assist these households because they do not receive a bill directly from the utility. Furthermore, assistance programs that focus only on direct customers in single-family homes may not be reaching a significant portion of low-income households in the service area.

Utilities contemplating assistance to non-bill-receiving customers are confronted with a complex set of challenges, including legal and administrative ones surrounding assistance to non-customers. Direct assistance to customers through means such as discounts to landlord accounts and vouchers presents several issues, including:

- Difficulties with effective communications and engagement to non-bill-receiving customers and their landlords
- Inability to cost-effectively obtain eligibility data, because the residents have no direct relationship with the utility
- Prohibitions to distributing funds to non-customers

- Low participation rates for landlords and concerns over recordkeeping requirements
- Concern that landlords cannot be guaranteed to pass the rate discount along to the tenants
- Unintended negative consequences for low-income residents when vouchers, counted as income, result in disqualifying those residents from other critical aid programs such as food aid

An infrastructure-based approach to providing assistance to these indirect customers is for utilities to install submetering in multifamily dwellings and convert more tenant households into direct customers. Then the assistance program could offer the same tools to eligible renters as it does to low-income property owners. Unfortunately, retrofitting submetering into an existing structure is typically not cost effective for the utility or the landlord. Submetering could be installed during the construction of new housing, but making it a requirement could increase costs excessively and hinder local development.

Although indirect assistance poses similar communication and engagement challenges as direct assistance does, it raises far fewer operational and legal issues and can provide appreciable benefits to low-income residents by supporting and stabilizing their financial circumstances. Some options for indirect approaches include supporting low-income households in applying for and obtaining available assistance, and supporting households through partnerships with local organizations. The WRF study stresses that the benefits of providing support to low-income customers' financial circumstances are well documented (Morin, 2017).

Critical to the success of either direct or indirect assistance to hard-to-reach customers is effective partnerships with organizations that engage with and assist low-income residents. These organizations have contacts and implementation experience that utilities often do not have. Local partnerships represent an opportunity for the organizations and the utility to provide new tools and support to residents in need. While effective approaches for assisting low-income, hard-to-reach residents is still not widespread, utilities have turned concerted attention to assisting this group.

4.5 Small Businesses

Small and local businesses may also be offered assistance by some utilities. Criteria for qualification as "small businesses" may include measures such as total revenue and number of employees. Utilities may be able to rely on another organization's certification to qualify businesses for assistance.

As with seniors or low-income customers with wealth, not all small businesses are making critical trade-offs in order to pay their wastewater bills. Arguments in favor of targeting small businesses for assistance recognize that small businesses may operate on a margin that does not allow them to weather some challenges that are easier for larger businesses to withstand.

As the introduction to this section mentioned, however, assistance programs must be developed within legal requirements, and with attention and care so that cost-of-service principles are not undermined, threatening the defensibility of the user charge mechanism. Selecting small businesses or all business for assistance can present challenges for legality and defensibility of rate structures.

5.0 STAKEHOLDER PERSPECTIVES

The design and implementation of CAPs, rate structure, or operational approaches to affordability, as with other utility initiatives, should be developed with stakeholder engagement and outreach. This section provides some broad perspectives and common points of view on affordability programs that utility policymakers may encounter. Naturally, every community has specific concerns and circumstances that affect stakeholders' reactions to and perspectives on the utility's approach to affordability. Each utility must gather its own data from stakeholders and develop an engagement plan that fits its community and goals.

Customer assistance programs and some rate structure–based approaches to affordability involve changes in the distribution of cost recovery among customers so that utilities can maintain their overall revenues while providing vital service and assistance to certain customers. Such changes in cost recovery can only be avoided when the program is funded via charitable donations and non-rate revenues, or the utility achieves affordability via cost-management efforts. Thus, affordability programs may generate conflict among stakeholders.

Perspectives vary on the overall importance of affordability and the means by which it should be achieved. Some regard clean, safe, and affordable water as a fundamental human right, while others may see it as less central—and perhaps only as a means for improving utility bill–collection rates and administrative efficiencies by reducing account delinquencies. Parties may disagree on what constitutes affordable water and wastewater service. With respect to achieving affordability, stakeholders may disagree on whether assistance and rate structure–based approaches are in conflict with cost-of-service utility pricing and, if they are in conflict, whether this indicates that a radical departure from such pricing models is warranted.

Some stakeholders feel that affordability of wastewater service should be addressed through federal action, whereas others see it as a state or local solution set.

Affordability programs are more likely to succeed if utility managers can anticipate and respond to all stakeholders' concerns. Such stakeholders include:

- Residential ratepayers
- Nonresidential ratepayers
- Advocacy groups
- Government regulators
- Local elected leaders
- Utility managers

5.1 Residential Ratepayers

Although some residential customers pay little attention to utility bills, others notice and are affected by any change in rates and rate structures. The current cost of utility bills as well as the rate structure influences the response of customers to rate changes. Low-income customers who face the threat of water shutoffs will appreciate affordability programs, especially those with easy steps for enrollment. Alternatively, if ratepayers who do not need an affordability program have to pay the financial burden, there might be some opposition. Multiple factors including the current rate structure and costs, recent rate increases, and the current political climate will influence customers' response to an affordability program. Stakeholder engagement and public outreach highlighting the importance of water affordability can help to build community support among customers who do not directly benefit.

5.2 Nonresidential Ratepayers

Like residential ratepayers, small nonresidential ratepayers (e.g., small commercial and industrial users) may or may not pay attention to or be burdened by utility rates. However, large commercial and industrial customers—for whom a small rate increase may translate into a much larger monthly expense—tend to notice any policy change that disproportionately affects their bottom lines. Companies that are invested in the local community may be willing to pay an increased price for equitable water access, especially if utility leaders and advocates can clearly illustrate the need for an affordability program. However, other nonresidential ratepayers may prescribe to strict cost-of-service principles or may also be struggling financially and not be willing or able to contribute to the cost of customer affordability programs. Others may weigh the negative effects on their profits versus the improvement in their employees' quality of life.

5.3 Advocacy Groups

Advocates for low-income and historically disadvantaged community members, and specifically for affordable wastewater services, will stress the essential nature of all water services. They view the need for assistance as immediate and a matter of human rights. They also tend to view unaffordable water as connected to or symptomatic of historic and pervasive discrimination.

5.4 Government Regulators

Government regulators tend to focus on the ability of utilities to maintain financial capabilities to meet current and future regulations. Utilities should prepare to address regulators' concerns about how lowering the costs paid by certain customers will affect the revenue of the utility and its financial obligations to maintain regulatory compliance. Regulators should be supportive of the utility providing affordable service to its customer base as long as the utility complies with permits and regulations.

5.5 Elected Leaders

Elected leaders, specifically local leaders, are sensitive to how residents and major industries are affected by changes in their utility bills. They may weigh the political and community benefits of an affordability program based on ratepayer support and opposition. They may find it important to help ensure that nonresidential customers that benefit the community are not financially overburdened, and some may consider access to water as an essential service. Elected leaders may be concerned about equity among members of their community and might be aligned with advocacy organizations that are pushing utilities to ensure all people have access to affordable water.

5.6 Utility Managers

While utility managers typically understand their responsibility to supply clean, safe, and reliable water and wastewater services, some may have different perspectives on the responsibility to address customer affordability, and they are also challenged by the increased planning and costs required for adopting and administering CAPs. Utility managers need support in preparing and executing stakeholder engagement programs. Some utility managers might choose to focus on the drawbacks to offering CAPs, and others may feel strongly that the utility should play a key role in assisting customers. Learning from the best practices of other utilities is one way that utility managers might lower the burden of developing affordability programs and help them reach a consensus before presenting an affordability approach to ratepayers.

6.0 CONCLUSIONS

The affordability program options discussed in this chapter provide tools for utilities, regulators, and community stakeholders to substantively help address their communities' customer affordability challenges. Utilities have different perspectives on their role in addressing affordability, and there is no customer assistance strategy or program that works best for every utility. However, given that capital and operating costs are expected to continue to rise, causing most utilities to face additional affordability challenges in the future, the tools discussed in this chapter can provide a menu of options that utilities can use to help address their objectives.

7.0 REFERENCES

American Water Works Association. (2018, October 24). AWWA policy statement on affordability. https://www.awwa.org/Policy-Advocacy/AWWA-Policy-Statements/Affordability

Giardina, R. (2019, March 7). A twist on cost of service. AWWA-WEF Utility Management Conference, Nashville, TN.

Morin, I. (2017, November). *Customer assistance programs for multifamily residential and other hard-to-reach customers.* Water Research Foundation.

Rothstein, E., Berahzer, S. I., Crea, J., & Matichich, M. (2021, September). Affordability and equity considerations in rate-setting. *Journal AWWA, 113*(7), 36–47. https://doi.org/10.1002/awwa.1766

Stratus Consulting. (2013). *Affordability assessment tool for federal water mandates.* U.S. Conference of Mayors, American Water Works Association, and Water Environment Federation. https://www.awwa.org/Portals/0/AWWA/ETS/Resources/AffordabilityAssessmentTool.pdf

Teodoro, M. (2018, September). Inoculation & therapy: Why rate structures, not assistance programs, offer the most promising path to water affordability. https://mannyteodoro.com/?p=610

University of North Carolina Environmental Finance Center. (2017, July). *Navigating legal pathways to rate-funded customer assistance programs: A guide for water and wastewater utilities.* Association of

Metropolitan Water Agencies, Water Research Foundation, and Water Environment Federation. https://www.amwa.net/assets/PathwaysRate-FundedCAPs_July2017.pdf

U.S. Department of Agriculture. (n.d.). Water and environmental programs. Rural Development. https://www.rd.usda.gov/programs-services/water-environmental-programs

U.S. Environmental Protection Agency. (2016, April). *Drinking water and wastewater utility customer assistance programs.* https://www.epa.gov/sites/default/files/2016-04/documents/dw-ww_utilities_cap_combined_508.pdf

U.S. Environmental Protection Agency. (2020). EJSCREEN: Environmental Justice Screening and Mapping Tool. https://www.epa.gov/ejscreen

4

Funding Options

The diverse array of approaches to addressing affordability, as discussed in Chapter 3, have a similarly broad spectrum of financial implications, particularly in the ways that affordability program costs are funded (and allocated across customer classes). External funding sources may include various governmental entities and philanthropic and community organizations; utility funding may be delivered through rate-setting practices or programmatically through such measures as customer assistance programs (CAPs).

The ways that affordability measures are funded, and how and why selected funding sources are constrained, have profound implications for designing affordability measures best suited for a specific community. For example, utility revenue funding options that may reliably deliver funding for CAPs in urban settings may be impractical for small rural systems with limited size and diversity of customer populations. In these cases, external funding sources may be critical for ensuring access and affordability of basic services. Because of the diversity of potential funding options and their implications, a particular challenge for utility service providers is defining a funding strategy that leverages available options and complements other organizational objectives.

1.0 FUNDING OPTION ATTRIBUTES

In defining a funding strategy, an array of different attributes may be considered, ranging from legal constraints on the ways funds are distributed to the administrative burdens imposed to satisfy funding provider requirements. In general, from a utility's perspective, funding sources may be categorized into three broad categories, within which there are a variety of individual options and programs:

1. External nongovernmental funding
2. Federal, state, and local government funding sources
3. Internal funding

These differences in funding sources impose a number of different constraints to consider in shaping a cogent and sustainable funding strategy. Paramount among these is the evolving legal framework defining each community partner's respective freedom to act.

1.1 Legal Considerations

The means by which funds may be secured and distributed to economically disadvantaged utility customers is evolving with recognition of water affordability challenges. Historically, wastewater services (like drinking water services) have been viewed largely as delivery of a commodity service where instances of nonpayment are addressed through collection measures, including service interruptions and placement of liens. Under this construct, measures to ensure access and affordability of service for low-income customers may be cast as subsidies, subject to legal constraints that limit or preclude entirely use of utility revenues. Rate relief or programmatic measures, like CAPs, may be vulnerable to challenge on the basis that discounted rates or assistance program funding amounts to discriminatory rate-setting practice. These barriers with respect to CAPs were cataloged in a nationwide review of legal constraints that summarized the legal landscape: "Ambiguous and restrictive statutory language has created the perception in many states that utilities are not allowed to tap their primary revenue source (customer rate revenues) to fund these programs, leaving utilities to have to find and develop other sources of funding" (University of North Carolina Environmental Finance Center, 2017, p. 7).

Yet, there is growing recognition that water and wastewater services deliver profound benefits that are not measured through commodity meters. Many people in the water sector believe that universal access to safe, reliable, and affordable water and wastewater services is a human right that addresses public health and sanitation needs fundamental to modern society. While these benefits are not necessarily a function of individual customers' water usage characteristics, they are nevertheless a fundamental attribute of providing water and wastewater service today, as amplified by recent experience with the COVID-19 pandemic (Rothstein et al., 2021).

Redefinition of the historical construct by which water and wastewater services are regarded to recognize the importance of ensuring universal access and affordability has the potential to lower historical barriers to funding affordability measures through utility revenue. In doing so, utilities' approaches to determining revenue requirements, allocating cost responsibilities, and designing utility rates may be reconfigured to address water affordability as a matter of course rather than as an afterthought. Progressive rate design and utility funding of CAPs may be aligned to regulatory compliance requirements and cost-of-service standards rather than subject to challenge, as was discussed in Chapter 3.

1.2 Eligibility Considerations

Legal considerations also directly influence the populations for whom water affordability measures are rendered, as these considerations often largely dictate customer eligibility for receipt of assistance. External funding through philanthropic giving may be relatively boundless, subject to the preferences and determinations of private donors, while federal and state governmental programs are subject to legislatively prescribed eligibility limitations. Most commonly, eligibility is defined by reference to household income measures. For example, most CAPs provide funding for utility customers who can demonstrate that their household incomes fall at or below defined income thresholds (e.g., 150% of the federal poverty line). Some communities and utilities have also elected to designate eligibility or prioritize selected customer groups—including seniors, veterans, and others—as a matter of local policy.

Two related and particularly important considerations in determining eligibility criteria are (a) the extent to which the defined criteria align with other social service programs designed to address economic disadvantage, and (b) the requirements placed on potential assistance recipients to evidence their eligibility and receive assistance. In general, utility service providers are not well positioned to screen customer documentation and determine income eligibility, as this has not been a utility customer service function or core competency. Utilities may partner with local social service agencies with established administrative procedures, yet must be mindful of the administrative (and personal) burdens placed on recipient customers if objectives to achieve meaningful levels of program participation are to be realized.

1.3 Program Features

Wastewater and water affordability funding may also be dictated based on the specific features of the programs or practices involved. Customer assistance programs that include delivery of water efficiency measures to help customers manage their water use and future bills may facilitate utility funding, whereas programs focused solely on bill assistance may be vulnerable to challenge. Similarly, rate-setting practices that segregate low-income customers for special rate treatment may be more subject to challenge than generally applicable rate designs that are structured to address water affordability concerns.

As with eligibility, the source of funding may be deterministic. Where external funding sources may largely define program features and target particular sources of concern—like emergency relief or housing stock deficiencies—federal and state program funding will typically be restricted to measures enumerated in authorizing legislation. In tailoring program components, utility service providers and potential community partners may benefit from researching the specific attributes of water affordability challenges in the communities they serve. Housing stock characteristics, income distributions, renter versus owner status, household sizes, water usage patterns, and social service program participation rates are but a few of the program design considerations that may vary dramatically across utility service areas.

1.4 Administrative/Monitoring and Reporting Requirements

Additionally, service providers and community partners must be mindful of the different administrative, data collection, and reporting needs associated with different measures intended to address water affordability. Water and wastewater rate designs structured to ensure the affordability of service levels required to meet basic human health and sanitary needs warrant evaluation of the extent to which billings to the economically disadvantaged have been moderated, and the incidence of payment delinquencies minimized. Careful tracking of the receipt, distribution, and use of CAP funds is important to ensure program effectiveness and the presence of adequate controls. Federal and state government funding sources typically require extensive program monitoring, auditing, and reporting to ensure appropriations meet legislative intent. Data requirements, collection procedures, and reporting protocols may vary dramatically, often depending on funding source. For example, federal program compliance reporting may be substantively different from that required by philanthropic organizations and community partners. Yet, collection of basic information on service area attributes affecting water affordability, and information regarding if and how relief has been rendered (e.g., participation rates, service interruptions and payment delinquencies reduced) is essential to improving and refining collective understanding of water affordability challenges.

1.5 Funding Adequacy and Stability

Fundamentally, service area attributes and affordability statistics are critical for assessing funding needs and the adequacy of funding for different forms of assistance—both of which, in turn, may shape a funding strategy. While limitations of federal and state funding prevail—traditionally in the form of infrastructure grants and loans, and more recently with the federal Low Income Water Customer Assistance Program—local service providers and community partners may only meaningfully affect water affordability challenges by establishing some sense of the scope of community-specific needs. Challenges in urban settings, often with geographic concentrations of endemic poverty, present substantively different needs and funding options than suburban or rural areas where both the prevalence of poverty and available resources are more (at least geographically) distributed. Water use, billings, and payment patterns may differ substantially across economically disadvantaged property owners and renters subject to different collection practices.

In practice, funding levels for water affordability measures, whether through external contributions and programs, governmental programs, or allocations of utility revenue requirements, will generally not be sufficient to eliminate low-income customer challenges. Water affordability issues are symptomatic of larger societal issues stemming from the incidence of endemic poverty. However, an integrated funding strategy that leverages available resources may be developed to render meaningful relief and sideline the use of punitive collection practices while safeguarding financial integrity. As noted, these available resources may be grouped into two broad categories: external funding sources, and internal utility funding sources.

2.0 EXTERNAL FUNDING SOURCES

Perhaps the most common sources of funding for wastewater and water affordability measures involve entities external to the utility services providers, reflecting (at least perceived) historical barriers to the use of utility funds for this purpose. These external funding sources may include (a) state and local general government social services agencies, (b) other utilities, (c) philanthropic organizations, (d) faith-based and community organizations, and (e) federal funding. Across and within these categories of external funding sources are different parameters that affect program structure, scope, and the attributes outlined above.

2.1 General (State/Local) Government Social Service Agencies

Established social service programs (including selected housing assistance programs administered by the U.S. Department of Housing and Urban Development and Temporary Assistance for Needy Families [TANF] and Low Income Home Energy Assistance Program [LIHEAP] administered by the U.S. Department of Health and Human Services) have often referenced utility bills among the essential needs to be addressed in providing relief to qualifying low-income households. Historically, energy (electric, gas) bills have been of primary concern, and program attributes have reflected this focus. For example, assessments of funding needs and allocations of federal dollars have prioritized relief for low-income households facing home heating needs in cold-weather states. Further, program features have often included provision of home energy audits. In general, the legal framework for these programs, and specifically eligibility criteria, are dictated by legislative provisions, which also largely determine administrative and reporting protocols.

Wastewater services providers may have the opportunity to partner with their local social services agencies to coordinate their programs' outreach, eligibility screening, and delivery of assistance measures but, at a minimum, should be knowledgeable about these program offerings and positioned to provide referrals.

Alternatively, some city or county governments could earmark a portion of tax revenues to fund their wastewater utility's CAPs aimed at improving customer affordability. Under this option, utilities could complete the required analysis to determine the funding levels needed to deliver different forms and levels of assistance and develop their customer assistance program. A general fund could support the program, at least partially, with tax revenue and thereby obviate potential rate challenges. Given the competing priorities of local governments, the ability and willingness to dedicate tax revenues to support water and wastewater affordability measures may be difficult to secure, especially as utilities are generally rate revenue supported.

2.2 Other Utilities

Many electric and gas utilities have established low-income CAPs partly because low-income energy assistance programs have been in place for at least 40 years (with the establishment of LIHEAP (through Title XXVI of the Omnibus Budget Reconciliation Act of 1981). These utilities have often established partnerships with local community organizations and social service agencies to conduct community outreach, screen applicants for eligibility, and coordinate program delivery. By aligning eligibility requirements with a utility's established low-income assistance program, wastewater and water affordability programs may limit the administrative burdens placed on program applicants to evidence eligibility. By coordinating administration and delivery, overall efficiency and effectiveness of assistance measures may be enhanced. In doing so, some program features may be accentuated. For example, many energy utility assistance programs feature home energy audits for participating households, often including installation of low-cost energy-saving measures. Performing a water use efficiency review along with the home energy audit can, in many cases, provide substantial and longer-term assistance to participating households than that delivered with limited-term bill assistance. Efficiencies in program outreach, administration, monitoring, and reporting may all be realized via coordination across utility service providers in collaboration with local community organizations. These efficiencies may enable individual communities to begin addressing chronic inadequacies in the levels of funding available to meet low-income utility assistance needs.

2.3 Philanthropic, Faith-Based, and Community Organizations

On the other end of the spectrum in terms of funding stability (and generally adequacy) are a range of programs delivered through philanthropic, faith-based, and community organizations. These organizations rely on donations from their memberships or fund-raising activities and, as a consequence, are typically subject to acute funding instability. Economic malaise that may exacerbate wastewater and water affordability challenges often affects charitable giving such that less funding may be available when needed most. Eligibility for these programs is generally a function of the dictates of the delivering organizations and their community partners, and therefore may be focused on, for example, delivering assistance to seniors, veterans, or church members. As such, these programs may not align in terms of eligibility requirements or program features with CAPs established through legislation or regulatory measures. Moreover, these programs are subject to different financial auditing and performance monitoring requirements such that evaluating program effectiveness and efficiency—both individually and collectively—is particularly challenging.

2.4 Federal/State Funding Sources

In light of the limitations of external funding described above along with growing affordability challenges across the nation, selected federal and state programs have been instituted and can potentially pilot larger federal and state programs (support for which has been galvanized by the COVID-19 pandemic). For example:

- In 2015, California passed legislation directing the State Water Resources Control Board to prepare a plan by 2018 to cover funding and implementation of the Low-Income Water Rate Assistance Program. The legislation is termed "California's Human Right to Water Policy" and states that "every human being has the right to safe, clean, affordable, and accessible water adequate for human consumption." The board has recommended the use of progressive revenue sources (i.e., taxes) to fund most of the program costs to avoid burdening the very state residents the program seeks to serve. The board has also recommended direct-bill credits with three tiers of charges; household eligibility is based on incomes below 200% of the federal poverty level (California State Water Resources Control Board, 2020; California Water Boards, 2017).
- In 2020, a federal program was funded through U.S. Congress appropriations of $638 million under the Consolidated Appropriations Act of 2021 and an additional $500 million under the American Rescue Plan Act of 2021. The Low-Income Household Water Assistance Program (LIHWAP) provides funds to assist low-income households with water and wastewater bills. LIHWAP grants were made available to entities (states, the District of Columbia, Puerto Rico, other U.S. territories, and federal- and state-recognized Indian tribes and tribal organizations) that received fiscal year 2021 LIHEAP grant funding. Under the program, grantees are required to provide funds to owners or operators of public water systems or wastewater treatment facilities to reduce arrearages of, and rates charged to, households for such services. The funds can be used to address arrearages, reconnection and late fees, and reductions in rate charges. The Low-Income Household Water Assistance Program is administered by the U.S. Department of Health and Human Services. However, there is no permanent authorization for LIHWAP (Office of Community Services, 2021).

Initial designs of these federal and state programs generally contemplate a mixture of direct-bill assistance to eligible households through utility bill credits and income tax credits for renter households. These programs, initially developed as a component of the COVID-19 pandemic response, evolved in the context of a general moratoria of service shutoffs for nonpayment. Program eligibility criteria, features, and administrative processes continue to evolve with changing federal and state roles in addressing water affordability.

3.0 UTILITY FUNDING SOURCES

For many larger (generally urban and suburban) community water systems, these external funding sources—whether private or public—may complement utility revenue funding. (As noted in Chapter 3, the legal framework for use of utility revenue funding was documented in *Navigating Legal Pathways to Rate-Funded Customer Assistance Programs* [University of North Carolina Environmental Finance Center, 2017]. This legal framework has continued to evolve, and generally has become more permissive of utility revenue funding, particularly given heightened water access and affordability concerns raised by the COVID-19 pandemic.) These larger community water systems typically have more resources to help address affordability issues internally, and funding of water affordability measures—potentially including progressive rate designs, direct assistance programs, and supporting water use efficiency programs—are included in system revenue requirements (rather than disassociated with annual calculations of rate and charges). In this case, significant questions arise as to how these revenue requirements will be allocated across customer classes and affect the structure of utility rates and charges (for more details, see Rothstein et al., 2021). For example, related costs may be allocated to the general residential class or, perhaps most compellingly, on a "common-to-all" basis reflecting collective public health benefits. In some cases, utilities may elect to highlight funding through different forms of surcharges, or simply recover costs through general customer or volumetric charges.

While utilities have noteworthy discretion to tailor rate designs and establish eligibility criteria for programmatic measures, as major economic actors in the communities they serve, coordination with other local social service and poverty assistance providers is advisable. This may be particularly helpful with outreach efforts to (understandably wary) disadvantaged community members. Utilities may also structure their affordability efforts to align with other imperatives and community values when defining program features. For example, many utilities have implemented assistance programs that provide water use efficiency assistance to complement their ongoing water conservation programs. Others have tailored program offerings to render assistance to seniors, veterans, or those facing health and/or employment crises.

For those utilities with adequate economies of scale in service delivery, ensuring access to wastewater and water service and affordability may be viewed as fundamental components of the services delivered, and use rates as a primary mechanism to fund related costs. Unfortunately, for many other (generally smaller, rural, and tribal) communities, utility funding may be impractical because the related costs may not be readily redistributed.

4.0 POLICY AND REGULATORY IMPLICATIONS

The likely inadequacy of any one funding source underscores the importance of defining a mix of funding options to address the disparate needs of different communities. Options will continue to evolve as the water sector addresses wastewater and water affordability. If larger reforms related to poverty mitigation, income inequality, and environmental and social justice are made, these options may become more robust, and specific programs will present new opportunities to stitch together the frayed social safety net that is reflected in rising water access and affordability challenges.

5.0 REFERENCES

California State Water Resources Control Board. (2020, February). *Recommendations for implementation of a statewide low-income water rate assistance program*. https://www.waterboards.ca.gov/water_issues/programs/conservation_portal/assistance/docs/ab401_report.pdf

California Water Boards. (2017, February). *Affordable & safe drinking water initiative* [Fact sheet]. https://www.waterboards.ca.gov/water_issues/programs/hr2w/docs/data/fs020817_asdw_act.pdf

Office of Community Services. (2021, June). Low Income Household Water Assistance Program information session. Administration for Children & Families. https://www.acf.hhs.gov/ocs/training-technical-assistance/lihwap-training-resources

Rothstein, E., Berahzer, S. I., Crea, J., & Matichich, M. (2021, September). Affordability and equity considerations in rate-setting. *Journal AWWA, 113*(7), 36–47. https://doi.org/10.1002/awwa.1766

University of North Carolina Environmental Finance Center. (2017, July). *Navigating legal pathways to rate-funded customer assistance programs: A guide for water and wastewater utilities*. Association of Metropolitan Water Agencies, Water Research Foundation, and Water Environment Federation. https://www.amwa.net/assets/PathwaysRate-FundedCAPs_July2017.pdf

5

Affordability Program Effects on the Utility

As discussed in Chapter 4, affordability programs can require significant funding to meet their objectives. This funding may in part be derived internally by utilities: by increasing some ratepayers' charges if allowed through relevant statutes and regulations by public service commissions; or from other non-rate revenue sources such as voluntary contributions from customers to support affordability programs, revenues from interest earnings, sale of scrap materials, and delinquency fees. Available revenues that can be used for funding may limit a utility's ability to enhance existing low-income assistance programs or to create a new program. However, there are many benefits to implementing low-income assistance programs, and these benefits may offset the added costs associated with these programs.

Several benefit and cost considerations are presented in this chapter. A cost-benefit analysis may be a useful tool for affordability program decision-making, although completion of a purely quantitative cost-benefit assessment may be challenging for most utilities. Costs and benefits of programs will vary from utility to utility based on program considerations, specific utility costs, and program benefits. Benefits, depending on type, may be more qualitative than quantitative. However, even though challenging, performing such a cost-benefit analysis may be a useful effort.

1.0 BENEFITS

Some of the benefits of low-income assistance programs include:

- Alignment with environmental justice and social values
- Fewer delinquencies and write-offs
- Improved customer health and well-being
- Improved customer relations

These benefits are briefly discussed below.

1.1 Alignment With Environmental Justice and Social Values

The Water Environment Federation (WEF) and many clean water and drinking water agencies around the United States support the sustainable and universal access to clean water, drinking water, and sanitation. While utilities may have differing views on the utility's responsibility to address customer affordability issues, WEF supports upholding socioeconomic and environmental justice and equity at every level of utility planning, financing, design, construction, operation, and delivery of all water infrastructure and services (WEF, 2021). Furthermore, affordability program alignment with environmental justice and social values may not necessarily conflict with the definition of equity used in the context of setting utility rates. Equity as it relates to utility ratemaking has been foundational to utility rate-setting and has traditionally meant that utility rates follow a cost-based standard reflective of the cost incurred to serve the various types of customers within a utility's service area. Where utilities follow strict adherence to cost causation in rate-setting or where use of user fee revenues has been prohibited by state or local statutes or regulatory practice, some utilities have used non-rate revenues and external funding to support low-income customer assistance programs (CAPs).

While cost causation is recognized as a key consideration in rate-setting, it is often not the only consideration when setting utility rates; cost causation often needs to be balanced with other priorities and objectives, such as affordability, conservation, economic development, administrative ease, and simplicity. For decades, there has been an interest by many utility leaders and governing bodies to provide utility service that is affordable for low-income customers, and some utilities have justified deviating from strict reliance on cost causation as the basis for rate-setting to help ensure affordable water service for low-income customers. Many of these utilities have used user fee revenues to fund CAPs that provide bill discounts and other forms of assistance to low-income customers where state statutes have allowed the use of rate revenues to support such programs. See Chapter 4 for a description of internal and external funding options.

1.2 Fewer Delinquencies and Write-Offs

It is normal to write off some uncollectible bills as bad debt. Most utilities, especially larger ones, develop protocols for writing off bad debts. Effective managers closely monitor changes in the volume of bad-debt write-offs. A successful affordability program can enhance a customer's ability to pay bills on time and can help utilities reduce both the number of uncollectible bills and the average time of accounts receivable. The reduction in bad debts can help offset affordability program costs.

1.3 Improved Customer Health and Well-Being

While water sector services are fundamental to setting a national baseline for ensuring public health, safety, economic well-being, and environmental quality, they can only be provided if they are affordable for ratepayers. When economic hardship and a lack of affordability leads households to sacrifice other essential goods and services to pay their utility bills, these customers may sacrifice other life essentials, such as medically necessary prescriptions or doctor visits, healthy meals, or child care. In addition, the loss of water services because of the inability to pay may in turn result in the loss of housing. Affordability and CAPs can help reduce bill delinquencies and utility shutoffs for nonpayment and, in turn, improve the health and well-being of low-income households.

1.4 Improved Customer Relations

A well-conceptualized and well-implemented affordability program may improve the utility's public relations image. While many utilities may have maintained a good public image in the past without a formal affordability program, they most likely used considerable discretion in handling overdue accounts receivable. Now, formal affordability policies and programs can be an element of high-quality management standards designed to meet community needs as service costs escalate. When effectively communicated to policymakers and customers, a well-designed affordability program could help improve the utility's image. If customers consider the affordability program to be socially effective and cost efficient, then utility policymakers may receive more support from their constituents.

2.0 COSTS

In addition to utility benefits of affordability programs, there may also be significant costs. These costs can include direct costs, such as labor required to support administration of the programs, and some indirect costs, as discussed below.

2.1 Revenue Effects

Most wastewater and water system customers can afford their bills and pay them on time. The collection rates for utility systems typically average 97% or more for utilities that participate in industry surveys such as the National Association of Clean Water Agencies (NACWA) Financial Survey. However, utilities do normally have some uncollectible accounts that are written off as bad debts. Historically, such debts typically account for approximately 2% of a utility's projected annual revenues. Customer assistance programs that allow reductions in bills or other financial assistance for qualifying low-income customers may require utilities to forgo more revenues. Historically, affordability program participation has been low, but as a utility reaches more program participants, the financial effects of CAPs on a utility could increase. Determining which customers to offer assistance to, and how deep the discounts will be, can affect the level of revenue effects. For many existing programs that provide such bill reductions, the value of the discounts provided has often been in the range of 1.0% to 1.5% of total gross revenues from user fees. Providing such assistance can reduce the administrative costs of dealing with uncollectible accounts (the cost of disconnecting service and reinstating service after the overdue amounts have been reconciled either through payment or some other mechanism such as a payment plan). To determine the financial implications of such programs, utility staff should analyze the types of customers who qualify for assistance, the amount of assistance such customers typically need, and the cost of administering the program. The total program cost can thus be identified, projected, managed, and tracked by accounting staff.

2.2 Direct Expenditures for Addressing Affordability

Customer assistance programs for affordability can take a variety of forms including bill discounts, emergency assistance programs, conducting water audits, and providing help in addressing leaky plumbing to reduce water demands. In addition, emerging approaches for addressing affordability include more directly embedding affordability into the rate-setting process, for example, by identifying an initial block of payments that is based on the costs necessary to provide essential water and wastewater services to meet the basic human health needs of residential customers. Given this potential array of program elements to address affordability, the direct expenditures for a CAP will vary depending on a community's approach to addressing affordability, and may include:

- Staff time needed to administer the program
- Staff time or expense-outsourced costs to identify and qualify eligible customers, develop a database for customer classification and billing, coordinate the development of billing practices and protocols with rates and billing staff, discuss customer complaints and collection issues, and refer customers to other social service agencies when necessary
- Costs for conducting a cost-of-service study and implementation plan for moving to recommended new rate and charge programs, if the approach to addressing affordability includes embedding affordability-related considerations such as a "human health" factor directly into the cost allocation and rate-setting process
- Billing and administrative tools, as well as computer service and capabilities (e.g., computer programs, customer surveys, and access to public income and tax databases)
- Communication costs (e.g., educational programs, advertising, and bill stuffers)
- Staff time to evaluate program performance, develop metrics, regularly report on program performance, and design program revisions as appropriate

A number of wastewater and water utilities have found that it is both cost efficient and useful from a policy perspective to partner with agencies that are already in the business of qualifying residents for other low-income support programs. For example, the utility systems in both Cleveland, Ohio and Portland, Oregon partner with the social service agencies that help qualify customers for the federal Low Income Home Energy Assistance Program (LIHEAP), typically based on family size and the percentage that household income represents in relation to the federal poverty level. By partnering with these agencies, a utility can typically work through a contract arrangement to get the qualification process done at a lower cost than it would take for the utility to set up and staff its own separate qualification process. Partnering with these other agencies also helps to provide consistency within a community on how qualification for support is done, and it allows the utility staff to focus on the higher-level policy objectives for the affordability programs.

2.3 Reliability of Other Utility System Programs and Services

Affordability program costs, such as labor and expenses as described above, should be budgeted and fully funded as part of a utility's operating budget development at a level consistent with the utility's goals and objectives for the affordability program. The affordability program should not be viewed as competition for the budgets of other important utility system activities, such as ongoing operation and maintenance of system assets, financial planning, and implementing renewal and replacement of system assets as part of an asset management program. Each of these important utility system functions should be budgeted at levels required to achieve the planned level of service to customers and other system stakeholders.

2.4 Indirect Costs

Indirect assistance program expenses may include:

- Management and administrative staff time spent developing and maintaining the program, both by the utility's staff and in outreach and education programs with customers and other stakeholder groups
- Policymakers' time spent reviewing, discussing, and adopting any proposed program
- General administrative and operating costs
- The opportunity costs of resources devoted to assistance programs that otherwise could have been used elsewhere

2.5 Statutory, Regulatory, and Bond Covenant Limitations

Utility staff should be aware of any restrictions imposed by bond documents or other legal or regulatory commitments, and the statutory context within the state. Revenue bond covenants might restrict a utility's ability to waive any customer's costs. Also, the U.S. Environmental Protection Agency (U.S. EPA)'s user-charge regulations impose restrictions on waiving or lowering user charges for wastewater utilities regulated under the Clean Water Act (WEF, 2018). Many wastewater utilities are subject to these rules or may choose to comply with them de facto, so their staffs should seek appropriate legal counsel before implementing affordability programs. If the utility is subject to a state's rate regulations, the appropriate legal requirements and case law should be adhered to when implementing any affordability program. Studies such as *Navigating Legal Pathways to Rate-Funded Customer Assistance Programs* (University of North Carolina Environmental Finance Center, 2017) have found that there is a significant difference among the legal and regulatory contexts in the 50 U.S. states in terms of whether there is a clearly indicated permission or denial to utilize user fee revenues to fund CAPs.

3.0 REFERENCES

University of North Carolina Environmental Finance Center. (2017, July). *Navigating legal pathways to rate-funded customer assistance programs: A guide for water and wastewater utilities*. Association of Metropolitan Water Agencies, Water Research Foundation, and Water Environment Federation. https://www.amwa.net/assets/PathwaysRate-FundedCAPs_July2017.pdf

Water Environment Federation. (2018). *Financing and charges for wastewater systems* (5th edition, WEF Manual of Practice No. 27).

Water Environment Federation. (2021, April 15). Climate change position statement. https://www.wef.org/advocacy/position-statements/

6

Customer Assistance Program Administration

Many issues must be addressed before low-income customer assistance programs (CAPs) for water and wastewater utilities may be successfully implemented. However, in the rush to get programs up and running, these issues are overlooked or underestimated. As a result, programs often encounter problems that must be dealt with during and after implementation. This chapter identifies some of the most common issues encountered in program administration and discusses how they may be addressed early to produce the best results. Addressing issues and following the steps identified in Figure 6.1 can help ensure the success of a new CAP.

1.0 DEVELOPMENT OF POLICIES AND PROCEDURES

The first step in CAP administration should be receipt of explicit direction regarding the program's primary mission and objectives from the utility's

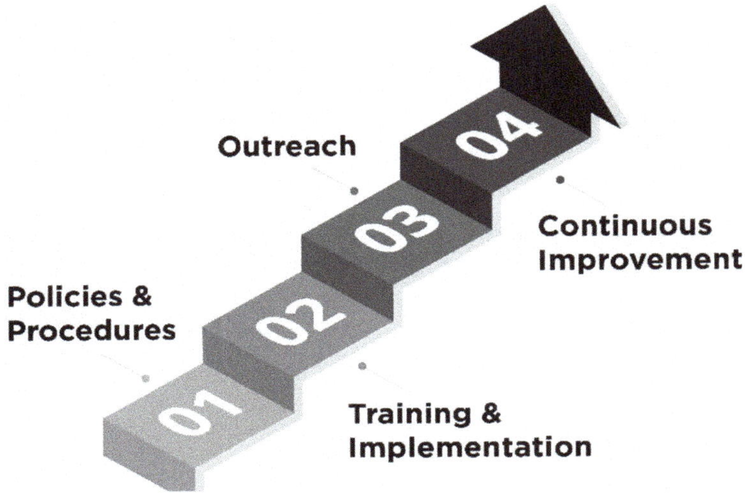

FIGURE 6.1 Steps to a Customer Assistance Program

decision-makers, such as elected officials or a governing board. From here, it is imperative to establish program-specific policies and procedures to serve as the foundation for long-term success. Some of the critical policy decision points are discussed in more detail below.

1.1 Program Qualification

The first major policy questions to be addressed revolve around program qualification.

- Who can receive assistance?
- How do they apply?
- What are the requirements to remain in the program?

1.1.1 Program Qualification Criteria

The first step to setting up a CAP is to determine qualification criteria. Some of the common criteria that are used for qualification include:

- **Account status.** Customer assistance must be connected to an active customer account.
- **Residency.** The applicant should demonstrate that the service address is their primary residence.

- **Customer type.** Assistance is often provided only to single-family, residential customers, but could be extended to multifamily customers in some instances.
- **Household income.** Income from all sources should be provided for each member of the household.
- **Number of household residents.** The number of household residents is essential for the calculation of assistance in many common measures such as the federal poverty level.

Each utility may ask for information and response verification however it chooses. However, there are some considerations to keep in mind. Identifying the minimum essential information simplifies the application process for both the applicant and the reviewer. Asking for some personal information, like Social Security numbers, may discourage fraud, but could also discourage applicants and increase privacy concerns. Finally, it is worth the time to investigate other local assistance programs and align qualification criteria to help streamline the application procedure for potential enrollees (see section 1.2, Program Interoperability).

1.1.2 Application Procedure

After deciding on qualification criteria, the assistance program should next develop the application procedure. The basic application asks for qualification criteria and an attestation that the applicant has provided accurate information. Some applications require acceptance of program responsibilities and verification of qualification criteria. The application should also provide clear instructions on how it can be completed. To aid accessibility, some utilities provide applications in multiple languages and allow applicants to fill them out online or download them from a website. Links to several CAP applications may be found in the case studies included in Chapter 8.

A review process must also be developed so that applications are reviewed in a consistent manner and the approval/denial result is both well documented and replicable. Application review may be done in-house by utility staff or outsourced to a contractor. In either case, the application procedure must be well conceived, and those administering applications must be well trained.

1.1.3 Requalification

Most CAPs require periodic requalification. The enrollee's economic circumstances may change, or the utility account/premise may change ownership. Periodic requalification assures enrollees that they are getting the appropriate benefit for their participation while also assuring the utility that no one is taking advantage of the program. Typically, CAPs require requalification every 12 to 24 months. Many programs provide flexibility on the requalification timeline to allow utilities to smooth out annual or seasonal spikes and ensure prompt application processing. For instance, if a new assistance program begins in July, the utility may plan for the first round of requalification to run from October through December of the following year to balance out the initial rush of reapplications and create a more even flow of work.

1.1.4 Program Removal

After considering how applicants qualify for a CAP, the utility must also consider what actions would trigger removal. Typically, program removal is immediate in the event of fraud, such as filing a false application, or for theft of service, such as meter tampering. Removal may also result for failure to requalify or failure to make timely payments.

1.2 Program Interoperability

When developing a new CAP, the utility should consider the benefits of interoperability with existing, locally available programs. For instance, mirroring the qualification criteria for an existing energy assistance program simplifies the application process. Providing an electric bill showing enrollment in the Low-Income Home Energy Assistance Program (LIHEAP) along with a water bill for the same service address and the same account holder would qualify the applicant automatically. In addition, a more streamlined application process may increase enrollment in both the energy and water assistance programs. Coordination among local assistance programs could be expanded to allow applicants to submit a single application for multiple programs. This "one-stop shop" for assistance could reach more of those in need and drive down administrative costs for the participating utilities.

1.3 Administration

The decision to administer a CAP internally depends on how each utility prioritizes and balances many factors. For example, Figure 6.2 illustrates that the desire to have more control over program details can adversely affect program cost or enrollment access.

1.3.1 Making the Case for In-House Program Administration

Many utilities are reluctant to outsource key functions, preferring to keep them in-house. One major advantage of internal administration is the ability

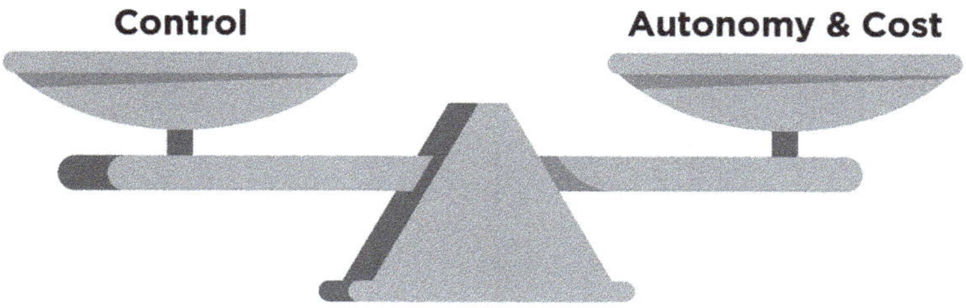

FIGURE 6.2 Balancing Cost Containment, Enrollment Access, and Program Control

to encourage departments within the organization to work together. A CAP will span many functions within the utility organization, including customer service, billing and collections, finance, communications, and information technology. If each of these functions report to the utility director, then they have both the responsibility and the authority to ensure the program meets its desired objectives. Typically, lead responsibility is assigned to one function, while the others coordinate to assist in design, development, and implementation. Depending on the size of the utility and the assistance program, a separate group may have to be constituted to provide ongoing administration.

Another advantage is the ability to maintain control over program decision-making. The utility that chooses to administer an assistance program in-house has the ability to make quick decisions. This ability allows the utility to set up its program more quickly or respond more effectively to changing circumstances. For instance, utilities with internally administered programs may have been able to act more decisively to adapt policies at the onset of the COVID-19 pandemic.

1.3.2 Making the Case for Contract Administration

There are often opportunities to outsource some, or all, of the administrative tasks of a CAP. Municipal governments often provide social services within areas that closely overlap the utility service area. In other cases, charitable organizations or 501(c)(3) organizations can be utilized at little or no cost to the utility. Finally, there may be opportunities to partner with electric or natural gas utilities that already administer their own assistance programs in the same geographic area.

Entities that specialize in providing social services within a community may be more efficient or effective in administering a CAP. Efficiencies stemming from economies of scale would allow existing social services providers to administer programs at a lower cost per enrollee. Contract partners could also bring years of experience in program administration that would not otherwise be available. Finally, contract partners could serve as a one-stop shop that provides multiple assistance offerings to needy customers, thus maximizing program interoperability.

1.4 Information Systems Considerations

Customer assistance program policies and procedures must be supported through the billing functions of the customer information system (CIS). At a minimum, the CIS must know which customers receive assistance and how to calculate their discounts. Systems considerations become more challenging if the CIS must calculate intricate discounts, track customer applications and recertification, and generate assistance program reporting. Finally, the CIS will have to interact with other information systems, such as the financial system to book discounted customer receivables and program costs. Each of these capabilities and system interactions must be mapped, built, and tested before the "go live" date of the assistance program.

2.0 TRAINING AND IMPLEMENTATION

The startup of a new assistance program is not unlike the startup of a new treatment facility process. Everything must be ready to run on day one, and must run seamlessly. Facility process startups ensure success through a careful system of testing and training, and assistance program startup should exercise the same rigor. Testing of data transfer between information technology systems confirms that applications flow to enrollments, which flow to accurate billing. Staff and contractor training confirms that manual processes occur in a timely manner and that customer service representatives are available and able to answer the inevitable wave of questions. As with a facility process startup, utilities should consider dry-run testing of the assistance program and have the full project team available to pitch in and troubleshoot on the "go live" date.

3.0 COMMUNITY OUTREACH

It is not enough to design and implement the ideal CAP. Stakeholders will insist that the utility undertake efforts to notify potential applicants of the program's existence and inform them how they can take advantage of its offerings. This is challenging because only about 30% to 50% of eligible customers typically enroll in water and wastewater utility assistance programs (Teodoro, 2021). Comprehensive community outreach programs help to assure stakeholders that the utility is making every effort to inform potential applicants about available assistance. Outreach methods include:

- Direct mail
- Bill stuffers
- Mass media and advertising
- Neighborhood organizations

Outreach efforts often take the form of frequently asked questions that inform potential applicants on how and where to apply for assistance, qualification criteria, and benefits they can receive. Outreach is most effective when it targets potential applicants where they live and work. Examples would include ads on public transit, or flyers for local social services providers.

A robust, well-subscribed CAP can help the utility secure needed rate increases. Affordability concerns are often cited as an obstacle in the rate-setting process. If the assistance program has a high subscription rate, governing bodies will be less concerned with affordability of service and more willing to increase rates to fund critical capital and operating costs.

4.0 CONTINUOUS IMPROVEMENT

Successful utility projects do not stop once the equipment is running. Pumps, tanks, and treatment processes require ongoing operation and maintenance to provide continued service. Customer assistance programs are similar in that they best serve customers when utilities prioritize continuous improvement. If metrics are not put in place and tracked, the ability to communicate program success will be adversely affected. Examples include:

- **Developing and tracking program performance metrics.** Are we enrolling eligible customers into the program? Are on-time payments increasing and customer shutoff rates decreasing?
- **Improving the customer experience.** Can the application procedure be simplified? Can we reduce processing time? How can we make applications more accessible?
- **Improving efficiency and reducing administration costs.** Can we spread applications more evenly across the year? Can we reduce the cost of

program administration on a cost-per-enrollee basis? Can we increase interoperability with other local assistance programs to reduce the application burden for customers and the application review burden for ourselves?

Developing and administering a CAP can be challenging for any utility. However, careful planning and execution will ensure the results meet program objectives in the most efficient manner.

5.0 REFERENCE

Teodoro, M. (2021, March 29). Batting .400: On the limits of means-tested assistance programs for water & sewer. https://mannyteodoro.com/?p=1856

7

Performance Measures

Appropriate performance measures enable utility decision-makers and the public to evaluate how well a customer assistance program (CAP) is performing and determine whether it is meeting its objectives. Performance measures can relate to the community at large, utility financial metrics, or be customer specific. Some measures are more insightful to explore when establishing a CAP, while others are more useful for understanding the program's effectiveness and efficiency.

1.0 INDUSTRY CONSIDERATIONS AND PERSPECTIVES

Well-established criteria for measuring a utility's financial capability that have been published by the U.S. Environmental Protection Agency (U.S. EPA) are often conflated with household affordability metrics. The criteria established by the U.S. EPA have been established relative to completing a program of capital project requirements aimed at regulatory compliance (traditionally eliminating combined sewer overflows). The criteria are structured to provide a broad measure of a community's ability to fund large capital projects and are not well suited to measuring affordability at the customer level, let alone the effectiveness of a CAP.

The U.S. EPA prepared a proposed update to the guidance it provides for measuring financial capability for public comment in 2021 and 2022, which included the key metrics shown in Table 7.1. The 2022 draft guidance builds on a U.S. EPA document first published in 1997 and is the source of a commonly used metric in evaluating affordability of wastewater service: a wastewater bill that exceeds 2% of median household income as a threshold for affordability. There are many flaws in using this metric as a stand-alone evaluation for the affordability of service for a community, which are identified in Chapter 2 and have been documented in several other sector publications. One of the most common critiques is that using a single point at the median income level does not offer meaningful insight into the affordability of service for all customers and especially low-income customers. The community measures do not provide any direct assessment of household affordability and should not be a focus for CAP measurement.

The household and poverty measures listed in Table 7.1 may be useful when establishing a CAP and can help inform appropriate thresholds for customer qualification and benefit levels. For example, imagine that the typical cost of a combined water and wastewater utility bill is $100 per month and that this represents 6% of the income for a household at the 20th percentile of income within the community. A utility may want to provide sufficient benefits to lower this typical bill so that it does not exceed a specific threshold, such as 4% for low-income households. This could be achieved with a $35 reduction in monthly costs, and the utility would have the flexibility to design the program elements accordingly.

The poverty rate and population that receives other public assistance can help inform how widespread affordability challenges are throughout

TABLE 7.1 Financial Capability Metrics, per U.S. EPA (2021, 2022)

Household Measures	Community Measures	Poverty Measures
Costs as a % of median income	Bond ratings	Upper limit of the lowest quintile income (LQI) vs. national LQI
Costs as a % of low income	Community debt leverage	Local income measure vs. national income
Typical customer bills	Unemployment rate	Population living under 200% of the federal poverty level
	Property tax revenue	Population on assistance
	Property tax collection rates	

a community. The U.S. Census Bureau publishes data that provides the number of households in a community at various multiples of the federal poverty level (100%, 125%, 200%, etc.), along with the number of households that qualify and receive other benefits. These are often part of the CAP qualification criteria and thus make good figures for assuming potential eligible populations. The following section introduces other measurements that could be used in lieu of the financial capability guidance when setting up, monitoring, and evaluating a CAP.

2.0 PERFORMANCE MEASURES WHEN ESTABLISHING A PROGRAM

When considering implementation of a CAP, it is important for a utility to understand customer behaviors, risks, and financial burdens. The measures presented in Table 7.2 can help decision-makers establish program parameters that provide an appropriate benefit to customers and can be used to estimate the cost of implementing the CAP. It is also important to know several key statistics about the utility, which will become points of comparison as the CAP is implemented and matures. The utility measures can be used to demonstrate the effectiveness of a CAP and whether the desired outcomes are being achieved. Performance measurement is an important aspect of continuous improvement when implementing a CAP, and it is generally good to monitor all measures as both nominal values as well as percentages.

As noted earlier in the chapter, the customer measures can be used to estimate the potential benefits offered customers, which will in turn allow for the calculation to estimate overall program cost. Table 7.3 presents a sample calculation that determines the potential cost of a CAP. Performing this calculation will require the analyst to make various assumptions for

TABLE 7.2 Example Baseline Affordability Measures (Before Establishing a CAP)

Customer Measures	Utility Measures
Number of delinquent accounts	Annual write-offs/bad debt
Disconnections or termination rate	Aged receivables
Bill amount and usage for delinquent accounts	Cost of collections
Bill amount and usage for typical accounts	Cost of customer assistance (if any)
Estimated eligible customers	Customer service caseload metrics

TABLE 7.3 CAP Cost Buildup Estimate

Estimated Annual Cost of Proposed CAP	
Annual income level of qualifying customers	$15,000.00
4% of annual income (affordability threshold)	$600.00
Estimated annual residential bill	$689.74
Annual subsidy required	$89.74
Projected qualifying customers	21,516
Estimated program cost (lost revenue)	$1,930,966
Additional administrative cost (10% of CAP)	$193,096
TOTAL COST	$2,124,062

participation rates. A range of values can be used to estimate the effects of low, medium, or high enrollment levels.

3.0 PERFORMANCE MEASURES TO EXAMINE AN EXISTING PROGRAM

It is important to monitor performance metrics after establishing a CAP, whether it be a robust program with many options for assistance or a more simplified structure. Performance metrics can be used to determine if targeted customer subgroups are being appropriately supported and if this assistance is translating into more stable financial performance for the utility overall. Selecting the right measures can also help utility managers understand what communication and outreach methods are successful at increasing participation rates, and if the programs enhance the perception of the utility within the community.

Table 7.4 presents additional performance measures that should be considered after implementing a CAP and are *in addition to* the measures presented in Table 7.2, which should continue to be monitored. Some of the measures below will not be applicable to every utility.

4.0 SAMPLE MEASUREMENTS

Collecting and tracking performance measures is most beneficial when the information is used to guide decision-making about ways to continuously improve a CAP. Utility managers may find it helpful to develop a basic report that shows the key metrics and results, and some will find it useful to translate the data into a visual dashboard that allows for quick understanding.

TABLE 7.4 Performance Measures After Establishing a CAP

Customer Measures	Utility Measures
Number of applicants	Number of CAP-related mailings
Percent of applicants approved	Number of CAP-related social media posts
Average benefit provided	Total benefit provided
Applicant customer types	
Bill amount and usage for enrolled accounts	
Payment plans created and completion rate	

A visual dashboard can also be useful when communicating to governing bodies or other stakeholders. In the sample graphics presented in Figure 7.1, utility managers and stakeholders can easily explore the CAP elements and effectiveness at achieving certain performance.

Utility managers need to understand if the implementation of a CAP has improved overall affordability of service as measured by key metrics. An effective CAP for a utility in Ohio may function differently than a program for a utility in Washington. Furthermore, the programs may have substantially different legal constraints, performance targets, or other policy considerations. Therefore, internal performance benchmarking is frequently the most useful form of performance measurement related to CAPs. If the desired performance is not demonstrated, the data may be used to indicate changes to the program that will improve performance.

5.0 LIMITATIONS AND OPPORTUNITIES

External performance measures, such as measures of poverty prevalence by comparing to national averages, may not accurately represent the real affordability burden placed on a household because of a water and wastewater utility bill. Two households in different regions living below the poverty line and receiving the same utility bill may experience different affordability challenges because of differences in the local cost of other essential services, and because of their individual circumstances such as differences in need for medical care. Utilities should look for sufficient data to guide their decision-making process without paralyzing this process with overwhelming data collection and analysis requirements.

Stakeholder engagement is essential to the success of a CAP. Customers cannot take advantage of a program if they are unaware of its existence.

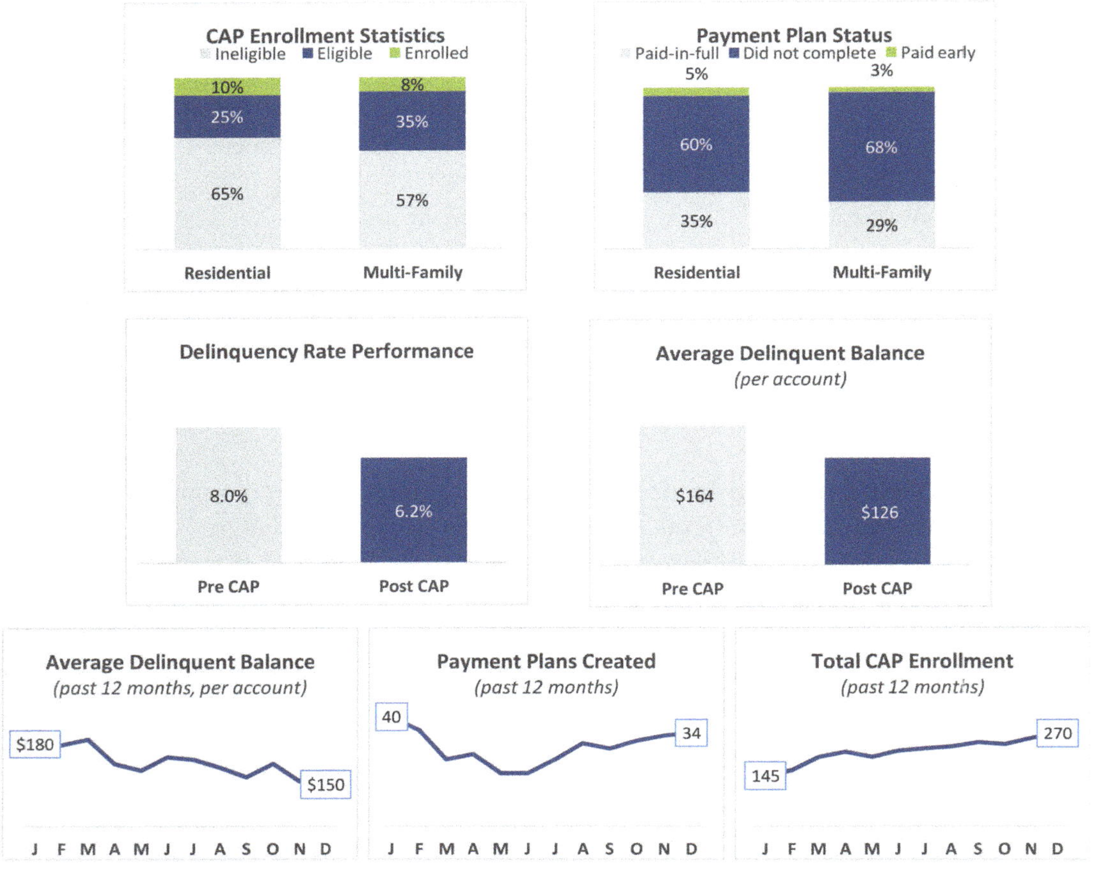

FIGURE 7.1 Sample CAP Performance Measurement Dashboard

Continuous and widespread community engagement will help drive enrollment and the benefits provided to customers. However, this effort requires staff resources and should be factored into the overall program costs. Tracking the success of various customer engagement initiatives will help a utility to become more efficient in its provision of the CAP over time.

6.0 CONCLUSIONS

Affordability means different things to different people (e.g., customers, utility managers, and elected officials), so even with an excellent set of measurement tools, different conclusions may be drawn about the program's effectiveness and efficiency. Nevertheless, utility managers and elected officials should establish performance measures because such tracking information can be used to assess progress and program success, and the performance measures could be supplemented by further inquiries, reports, and customer feedback to obtain a clearer picture of program effectiveness.

Managers should develop and implement performance measures for low-income assistance programs so they can evaluate the program and make strategic and tactical decisions about program activities and investments. Also, the public will want to know that utility officials are managing utility resources properly and prudently, and performance measures help provide such assurances. They also can help utility staff respond to questions about the program.

7.0 REFERENCES

U.S. Environmental Protection Agency. (1997). *Combined sewer overflows—Guidance for financial capability assessment and schedule development.* Office of Water, Office of Wastewater Management.

U.S. Environmental Protection Agency. (2021, January). *Financial capability assessment guidance.* https://www.epa.gov/sites/default/files/2021-01/documents/2021_fca_guidance_-_january_13_2021_final_prepub.pdf

U.S. Environmental Protection Agency. (2022, February). *Proposed 2022 clean water act financial capability assessment guidance.* https://www.epa.gov/system/files/documents/2022-02/2022-proposed-fca_feb-2022.pdf

8

Case Studies

This chapter provides several example case studies of customer assistance programs (CAPs) implemented by water and wastewater utilities to provide the reader with information on specific CAPs that have been developed and implemented. The case examples were selected to reflect utilities of different sizes, in different geographic areas, and with different program elements.

1.0 COLUMBUS WATER WORKS (GEORGIA)

The Columbus Water Works (CWW) provides drinking water and wastewater treatment to approximately 200,000 people throughout Muscogee County, Georgia. Residential water rates include a fixed monthly meter charge and a two-tier inclining block volumetric rate. Residential wastewater rates also include a fixed monthly service charge and volumetric rate based on the quantity of water used. The volumetric rate is capped at 42 475 L (1500 cu ft) for residential customers. Additionally, customers using more than 14 158 L (500 cu ft) per month are assessed a fixed combined sewer overflow charge per account.

The Columbus Water Works offers a Low-Income Credit Program to eligible customers that is composed of a fixed credit of $8.50 per monthly bill that can be applied toward a customer bill (CCW, n.d.). The credit is extended only to eligible residential customers who meet poverty guidelines based on household size. Customers who qualify for the city's reduced garbage rate are automatically enrolled in the low-income water and wastewater credit. Applications for the program are processed through a partnership with the city's 311 Citizen Services department. In addition to this program, CWW offers payment arrangements to customers and partners with the local United Way to offer other emergency assistance.

The Columbus Water Works has seen steady enrollment in its assistance program over the past five years. Increasing enrollment is a continued focus of utility management.

2.0 DC WATER AND SEWER AUTHORITY (DISTRICT OF COLUMBIA)

The DC Water and Sewer Authority (DC Water) provides retail water and wastewater service inside Washington, DC, and wholesale wastewater service in surrounding Maryland and Virginia counties. Retail customer bills include water and sewer volumetric rates; a metering fee and water system replacement fee fixed by meter size; an impervious area charge (CRIAC); and payment in lieu of tax, right-of-way, and stormwater fees that are remitted to the District of Columbia (District).

DC Water and the District government offer a host of assistance programs that target customer affordability challenges from many different angles. These programs include the following (DC Water, n.d.; Department of Energy & Environment, n.d.):

- The Residential Assistance Program or Emergency Residential Relief was established by the District in early 2021 to provide up to $2,000 in assistance to residential customers who have fallen behind on monthly bills during the COVID-19 public health emergency.
- The Customer Assistance Programs (CAP, CAP2, and CAP3) provide monthly bill discounts to income-qualified households. Qualification is based on household income and size. Applications are made through the District Department of Energy & Environment Utility Discount Program using federal Low-Income Home Energy Assistance Program (LIHEAP) criteria.
    - CAP provides a 100% discount on the first 11 327 L (400 cu ft) of monthly metered usage as well as a discount on the Water System Replacement Fee and CRIAC.

- CAP2 provides a 100% discount on the first 8495 L (300 cu ft) of monthly metered usage as well as a 50% discount on the CRIAC.
- CAP3 provides a 75% discount on the monthly CRIAC.

- The Multifamily Assistance Program provides discounts to eligible landlords who pass along savings to low-income tenants through reduced rents or homeowners dues.
- STAY DC is a District program that uses federal funds to provide DC renters and housing providers with rental assistance and help with utility bills, including water and wastewater.
- Nonprofit Relief assists eligible nonprofit organizations by providing a 90% discount on their monthly CRIAC.
- Extended Payment Plans help customers of any income level catch up on account delinquency.
- Serving People by Lending a Supporting Hand is funded by contributions and provides one-time emergency assistance to help customers maintain critical water and sewer service.

More information on District CAPs may be found on the DC Water website, dcwater.com.

3.0 DETROIT WATER AND SEWERAGE DEPARTMENT (MICHIGAN)

The Detroit Water and Sewerage Department (DWSD) serves more than 650,000 people, of whom approximately 40% have incomes below the federal poverty level (FPL). For decades, DWSD was challenged by high customer arrearages and bad-debt expenses resulting in accounts becoming eligible for shutoff. In 2016, new policies were put in place to help address the affordability issues in Detroit.

The Detroit Water and Sewerage Department implemented a Water Residential Assistance Program (WRAP) for residential customers who are at or below 200% of the FPL (City of Detroit Water and Sewerage Department, n.d.). The program freezes arrearages for 12 months, provides a $25 credit toward a customer's monthly bill for one year (a total credit of $300), offers up to $700 toward past-due balances to reduce frozen arrearages with the customer continuing payments under an established payment plan, provides water conservation audit services with up to $1,000 in minor home plumbing repairs, and connects customers with social services support, as needed.

The WRAP was developed as part of an agreement between DWSD and the Great Lakes Water Authority, which earmarks 0.5% of budgeted operating revenue annually toward the financial assistance program for the region over a 40-year period.

In addition, DWSD has implemented a 10/30/50 arrearage payment plan. Under this plan, customers who have past due balances or are delinquent on their accounts can spread their arrearage payments over a 24-month period with as little as 10% down payment on the past-due amount. If a participant misses a payment, they can re-enroll by making a payment of 30% of the remaining balance.

During the first year of program implementation, the program was successful in reducing customer disconnections from 44,000 to 13,000 accounts.

More information on DWSD's WRAP and its other customer assistance programs may be found on the department website, https://detroitmi.gov/departments/water-and-sewerage-department/dwsd-customer-service/customer-affordability-programs.

4.0 NORTHEAST OHIO REGIONAL SEWER DISTRICT (CLEVELAND, OHIO)

The Northeast Ohio Regional Sewer District (NEORSD or District) is the largest wastewater treatment provider in Ohio, serving 62 communities and nearly 1 million residents, and treating 340 billion L (90 billion gal) of wastewater each year. The Northeast Ohio Regional Sewer District charges a fixed monthly amount for wastewater service in addition to a wastewater usage charge based on water consumption. For stormwater services, residential customers are charged a tiered fixed monthly fee based on impervious surface area.

The Northeast Ohio Regional Sewer District offers several financial assistance programs that are designed to target different types of customers in need of assistance. This includes two bill discount programs that both provide a 40% discount on wastewater and stormwater bills for low-income customers. The Affordability Program, which is administered by NEORSD in partnership with a third-party nonprofit organization, provides this discount to households with incomes below 200% of the FPL and who own their home. The Homestead Program, which is administered by Cleveland Water, provides the same discount for customers with household incomes below $34,500 (for 2021), who are either at least 65 years old or receive disability, and who also own their home (NEORSD, n.d.). In addition to these programs, NEORSD offers crisis assistance that provides a significant discount on wastewater account balances for qualifying customers, as well as sewer and plumbing repairs for low-income households.

The Northeast Ohio Regional Sewer District's ongoing affordability efforts include a commitment to continued evaluation and adaptive management to ensure that the programs continue to achieve intended objectives and effectively address affordability challenges. In 2020, NEORSD evaluated the Affordability and Homestead programs toward this end (Clements & Sheridan, 2021). This evaluation found that participation in the long-standing Homestead Program was quite high, with an estimated 60% of eligible customers enrolled in the program (32,300 participating households). However, the Affordability Program had a lower enrollment rate of just over 11% (2,250 participating households), which is typical for the industry. In addition to current participation, the analysis also indicated that there were more than 20,000 customers within the service area who meet the income criteria for these programs and who have the wastewater/stormwater bill in their name but were not eligible because they do not own their own home. Additionally, NEORSD examined cost of living data for the Cleveland metropolitan area, finding that customers earning up to 250% of the FPL may still struggle to pay their water and sewer bills. By increasing the income threshold for the Affordability Program from 200% to 250%, an additional 21,000 customers would be eligible for assistance. As a result of these findings, NEORSD adopted the following potential program enhancements: (a) expanding the Affordability Program to renters who pay their own bills, and (b) increasing the qualifying income criteria. Expanding the program in these ways more than tripled the customers currently eligible for assistance.

In addition to the program evaluation, NEORSD recently worked with the third-party administrator of the Affordability Program to hire a "Water Champion," who is tasked with enhancing the District's robust customer outreach and enrollment efforts and further connecting at-risk populations to water and sewer bill assistance. This includes identifying and reducing barriers to participation in NEORSD programs for low-income customers. In implementing these changes to the program, NEORSD set a goal of increasing participation in the Affordability Program to at least 20% of eligible customers by 2026, adding an additional 10,000 accounts.

5.0 PHILADELPHIA WATER DEPARTMENT (PENNSYLVANIA)

The Philadelphia Water Department (PWD) provides drinking water, wastewater, and stormwater services to 1.58 million people in the greater Philadelphia region. Philadelphia's water and wastewater fixed charges are scaled based on the size of the meter serving the customer. Water usage rates are structured as a four-tiered declining block rate for all customer classes, while wastewater rates are charged on a uniform basis with all customers paying the same rates regardless of customer class. The stormwater structure includes fixed customer service charges, a charge per gross area, and a charge per impervious area. Single-family residential customers are all charged a flat fee based on the property's characteristics. Nonresidential properties are charged based on actual gross and impervious area measurements.

The Philadelphia Water Department offers the Tiered Assistance Program (TAP) for low-income customers, under which monthly bill payment amounts are determined based on a percentage of an individual customer's monthly income. Eligible customers must have incomes at or below 150% of the FPL or must demonstrate special financial hardship. As shown in Table 8.1, TAP caps participating customers' bills between 2% and 4% of their income based on FPL thresholds. For example, a TAP customer with income equal to 50% of FPL will receive a total bill fixed at 2% of their household income. For customers in this lowest income bracket, the minimum bill is $12 per month. The Tiered Assistance Program also allows participating customers to earn forgiveness of their pre-TAP debt after 24 full TAP payments. As of December 2020, more than 16,000 households were actively enrolled in this program (Philadelphia Department of Revenue, 2020).

In addition to TAP, Philadelphia also offers other targeted CAPs. The Senior Discount Program offers a discount of 25% for seniors more than 65 years old with incomes less than $32,300 per year (City of Philadelphia, n.d.). Conservation Assistance helps households below 150% of the FPL install water conservation devices and provides in-home education. Payment agreements and extended payment plans are available for customers behind on their bills; total bills plus arrears payments can be capped at 4% of income for households that earn between 151% and 250% of the FPL (below this, customers are eligible for TAP) (Philadelphia Department of Revenue, 2021). The Philadelphia Water Department also offers emergency loan programs to help customers avoid shutoffs, zero-interest loans for low-income homeowners to replace lead service lines, and free plumbing devices to help prevent basement flooding. Finally, in addition to its residential programs, the city also offers several programs to nonresidential customers, including payment arrangements and bill discounts.

TABLE 8.1 Philadelphia Customer Bill Maximum Levels

Household income	Maximum bill as % of income
0–50% FPL	2%
51–100% FPL	2.5%
101–150% FPL	3%
>150% FPL + Special Hardship	4%

6.0 SEATTLE PUBLIC UTILITIES (WASHINGTON)

Seattle Public Utilities (SPU) provides drinking water to approximately 1.5 million people living in the greater Seattle area, some through various wholesale suppliers. Seattle Public Utilities also provides wastewater and drainage services throughout the city to manage flooding and convey wastewater to King County for treatment. The residential water rates include a monthly base service charge and commodity rates based on the volume of water used. The commodity rate is a uniform rate during the off-peak season (September 16 through May 15); peak-season rates include a three-tier inclining block rate. The wastewater charge is entirely based on the average volume of water consumed during winter months (November through April). Drainage rates are a fixed charge assessed to each property based on the square footage of each parcel and are assessed as part of the property tax bill.

Seattle Public Utilities offers two primary CAP elements: the Utility Discount Program (UDP) and the Emergency Assistance Program. In addition to these programs, SPU offers payment plans to customers. The UDP provides bill assistance for seniors, persons with disabilities, and low-income customers whose household income is at or below 70% of the state median income. The program helps customers get current and stay current on utility payments by offering a discount of about 50% on the SPU bill (Seattle Public Utilities, n.d.-a). The income qualification is measured against an increasing scale based on the household size. Applications for the UDP can be made via an online enrollment form and may take up to two billing cycles to become active.

One unique feature of the UDP offered by SPU is an option that allows renters or other customers who do not directly receive a SPU bill to receive customer assistance. The City of Seattle also manages local energy bills through Seattle City Light. Customers who have their SPU services paid through a homeowner's association or landlord but still receive a Seattle City Light bill can receive fixed credits per month on their electric bill. The credit is calculated to provide about a 50% credit on the typical residential bill.

Seattle Public Utilities offers an Emergency Assistance Program for customers with more immediate needs who cannot wait the two billing cycles that the UDP may take to process. The Emergency Assistance Program is open to customers in single-family residences with gross household income below 80% of the state median income. Customers also must not have received emergency assistance in the current calendar year unless there are children under the age of 18 living in the household; those can receive assistance twice in a calendar year. Applications to the Emergency Assistance Program are made by calling customer service and can receive up to $461 toward their utility bill (Seattle Public Utilities, n.d.-b).

When enrollment was stagnant at about 14,000 households, SPU, Seattle City Light, and the Human Services Department embarked on an ambitious effort to double enrollment over the last 10 years by taking the following steps:

- Changed Seattle municipal code to allow all people in subsidized housing to enroll
- Lowered barriers to entry by moving from requiring three months of proof of income to just one month
- Allowed households in which the primary account holder is on Supplemental Nutrition Assistance Program (SNAP) to apply using SNAP verification
- Worked with housing providers to streamline the enrollment process and help new tenants enroll when they open their utility account
- Worked with the Seattle Housing Authority and the Washington Housing Finance Division to obtain a list of tenants who are potentially eligible for the program
- Targeted specific marketing to English as a Second Language customers

Seattle Public Utilities staff have found the most success when they make efforts to be inclusive of the people they are hoping to serve—not just listening to their feedback, but making efforts to implement it. For example, they have different members of the community test the application and provide feedback on the process. The utility's efforts to build partners in the community, lower the barriers to enrollment, and act on community feedback have led to a current enrollment of over 30,000 households, which is 50% of the estimated eligible population.

7.0 REFERENCES

City of Detroit Water and Sewerage Department. (n.d.). Customer affordability programs. https://detroitmi.gov/departments/water-and-sewerage-department/dwsd-customer-service/customer-affordability-programs

City of Philadelphia. (n.d.). Senior citizen water bill discount. https://www.phila.gov/services/water-gas-utilities/pay-or-dispute-a-water-bill/senior-citizen-water-bill-discount/

Clements, J., & Sheridan, C. (2021). *Affordability of wastewater services in the NEORSD service area and analysis of Affordability and Homestead Programs*. Northeast Ohio Regional Sewer District.

Columbus Water Works. (n.d.). Payment options. https://www.cwwga.org/customerservice/paymentoptions

DC Water. (n.d.). Customer assistance programs. https://www.dcwater.com/customer-assistance

Department of Energy & Environment. (n.d.). Receive discounts on your utility bills. District of Columbia. https://doee.dc.gov/udp

Northeast Ohio Regional Sewer District. (n.d.) Cost-saving programs. https://customerservice.neorsd.org/s/cost-saving-program

Philadelphia Department of Revenue. (2020). *Annual report to the mayor on the Tiered Assistance Program (TAP)*. https://www.phila.gov/media/20200715113740/Tiered-Assistance-Propram-TAP-annual-report-2019.pdf

Philadelphia Department of Revenue. (2021). Customer assistance program eligibility guidelines. https://www.phila.gov/media/20210217082849/Water-bill-customer-assistance-programs-eligibility-2021-guidelines.pdf

Seattle Public Utilities. (n.d.-a). Utility discount program. https://www.seattle.gov/utilities/your-services/discounts-and-incentives/utility-discount-program

Seattle Public Utilities. (n.d.-b). Emergency assistance program. https://www.seattle.gov/utilities/your-services/discounts-and-incentives/emergency-assistance-program

9

Future Considerations

Over the past several years, there has been a heightened focus on customer affordability in the wastewater and water sector because of the rising cost of utility service, changing demographics, efforts of the Water Environment Federation (WEF) and other sector organizations, and hardship issues highlighted during the COVID-19 pandemic. In addition, the amount of research, analysis, perspectives, and guidance on the topic of wastewater and water affordability has continued to increase and will continue to evolve over time. As of the time of writing this edition, several issues were either changing, under development, or simply identified as future research needs. These are summarized below.

1.0 U.S. ENVIRONMENTAL PROTECTION AGENCY FINANCIAL CAPABILITY ASSESSMENT

In January 2021, the U.S. Environmental Protection Agency (U.S. EPA) issued a proposed update to its 1997 *Combined Sewer Overflows—Guidance for*

Financial Capability Assessment and Schedule Development publication, and in February 2022, the U.S. EPA issued further refinements to its updated guidance for public comment. However, as of the time of this writing, this guidance has not been finalized and adopted by the agency. It is expected that this guidance will be adopted by the U.S. EPA in 2022 with possible further refinements.

2.0 AFFORDABILITY MEASURES FOR RESIDENTIAL CUSTOMERS

There are numerous metrics that have been developed to measure residential customer affordability. Some of these metrics capture the cost of other essential household needs or the local cost of living, such as the Low Income Housing Burden (available from the U.S. Census for some communities), the affordability ratio at the 20th percentile (Teodoro, 2018), and the MIT Living Wage (https://livingwage.mit.edu). These measures have limitations in their ability to be broadly applied to both utilities serving large urban or suburban communities and those serving small, rural communities, because of the lack of universally available local cost of living information. Developing affordability metrics that are broadly applicable and compare essential living costs with incomes, factoring in the local cost of living, is an area where future research and development could improve upon the affordability measures available today. In addition, future research into the prevalence and practices of low-income customers who defer some essential services for others (e.g., water service vs. health care) could help in measuring the effects of the lack of affordability of water and wastewater services.

3.0 AFFORDABILITY MEASURES FOR NONRESIDENTIAL CUSTOMERS

While the body of research, literature, and guidance pertaining to residential customer affordability continues to grow and expand, there has been little focus in the literature on affordability measurement and assessment for nonresidential customers (e.g., commercial, industrial, institutional, or nonprofit). Further research and analysis on nonresidential customer segments may be warranted to help provide a complete assessment of a utility's service area customer affordability.

4.0 STUDY OF WATER USAGE PATTERNS OF LOW-INCOME CUSTOMERS

There are conflicting perspectives in the water and wastewater sector regarding whether low-income customers use more or less water than the typical or median utility customer. Additional research and analysis of water usage by low-income customers could aid in quantifying the water sector bills of these customers and help to quantify the assistance that is needed.

5.0 FEDERAL WATER AND WASTEWATER FUNDING AND ASSISTANCE PROGRAMS

In response to the COVID-19 pandemic, the federal government enacted the Consolidated Appropriations Act of 2020 and the American Rescue Plan Act (ARPA) of 2021, which was intended to provide approximately $1.1 billion for public wastewater and water systems to reduce arrearages and bill effects to low-income households. These funds support the temporary Low-Income Household Water Assistance Program (LIHWAP) to provide relief to low-income households. As of this writing, it is not known whether LIHWAP will become a permanent federal assistance program, how it may change or evolve in the future, and the level of funding that it may provide. As this federal program—and potentially others—evolve, utilities will need to assess how local and federal programs can work in combination to provide the level of low-income customer assistance needed within a community.

More research is needed to develop the most efficient and effective ways to deliver federal funding assistance to the low-income customer recipient. This could entail leveraging the administration of other federal assistance programs, such as the Low-Income Home Energy Assistance Program (LIHEAP) and the Supplemental Nutrition Assistance Program (SNAP), to effectively deliver low-income assistance funding in the water sector. Further, additional research into the capacity requirements for administration of poverty assistance programs is not currently fully understood. Additional research on the capacity needed for community organizations or utilities to effectively administer customer assistance programs (CAPs) would help the wastewater and water sector better understand the administrative aspects of CAPs.

6.0 POTENTIAL CHANGES TO COST-OF-SERVICE AND RATE-SETTING PRACTICES

As described in Chapter 3, the practice of including a human health element of cost of service and cost allocation to help justify fair and equitable utility rates, while taking into consideration affordability, is being discussed in the wastewater and water sector. If the application of this approach were eventually to prove workable, it could have the potential to address some of the state legal limitations around rate-setting and funding existing customer assistance programs.

7.0 STATE LIMITATIONS ON FUNDING CUSTOMER ASSISTANCE PROGRAMS

As described in Chapter 4, ambiguous and restrictive state statutory language has created a challenge for utilities in many states regarding their ability to tap their primary revenue source (customer rate revenues) to fund customer assistance programs, resulting in utilities having to find and develop other sources of funding. As there is a growing recognition that wastewater and water services provide essential benefits to society, it is possible that some states will clarify or revise their statutory language in the future, thereby making it feasible for utilities in these states to use some portion of their rate revenues to fund CAPs.

8.0 REFERENCES

Teodoro, M. P. (2018). Measuring household affordability for water and sewer utilities. *Journal of the American Water Works Association*, *110*(1), 13–24.

U.S. Environmental Protection Agency. (1997). *Combined sewer overflows—Guidance for financial capability assessment and schedule development*. Office of Water, Office of Wastewater Management.

U.S. Environmental Protection Agency. (2021, January). *Financial capability assessment guidance*. https://www.epa.gov/sites/default/files/2021-01/documents/2021_fca_guidance_-_january_13_2021_final_prepub.pdf

U.S. Environmental Protection Agency. (2022, February). *Proposed 2022 clean water act financial capability assessment guidance*. https://www.epa.gov/system/files/documents/2022-02/2022-proposed-fca_feb-2022.pdf

Index

CPSIA information can be obtained
at www.ICGtesting.com
Printed in the USA
LVHW060604270522
719767LV00001B/5

9 781572 784093